The Future of Cloud

A Roadmap of Technology, Product, and Service Innovations for Telecoms

Thorsten Claus

Daniel Kellmereit

Yasmin Narielvala

Version 1.1

ISBN: 1450557414
EAN-13: 9781450557412
BISAC Category: Telecommunications
Reading Level: College Freshman
Search Keywords: Cloud Computing, SaaS, Innovation, Telecom, Virtualization

Abstract: This is not a Cloud Computing primer, nor does this book focus solely on the computing aspect of Cloud. We discuss the roadmap for technology and service innovations for telecoms in the medium to long term future. A strategic impact analysis leads to recommendations for potential value chain positioning. Through detailed interviews with telecom vendors, suppliers, customers, and venture capital firms we span an ecosystem of influencers and currents within the complex and evolving industries shaping the Cloud.

Printed in the United States of America.
10 9 8 7 6 5 4 3 2 1

Disclaimer

While the publisher and author have used their best effort in preparing this book, they make no representations or warranties with respect to the accuracy or completeness of the contents of this book and specifically disclaim any implied warranties of merchantability or fitness for a particular purpose. No warranty may be created or extended by sales representatives or written sales materials. The advice and strategies contained herein may not be suitable for your situation. You should consult with a professional where appropriate. Neither the publisher nor author shall be liable for any loss of profit or any other commercial damages, including but not limited to special, incidental, consequential, or other damages. No animals were harmed during the interviews.

The interviews in this book are the opinions of the persons involved and do not necessarily reflect the opinions of the parties' respective current or past employers. Furthermore, the interviews were taken at a specific point in time in the past. Circumstances, opinions, facts, products, services, and technologies might have changed since then. The interviewees might also have changed their position, responsibilities, or employer since then.

Trademarks

Photo Copyright

We would like to thank all interview participants for their insightful and thoughtful comments they shared with us: Jamie Allen, Simon Aspinal, Timo Bauer, Aneesh Chopra, Peter Coffee, Paul Curto, Ismael Ghalimi, Pete Grillo, Raju Gulabani, Tom Hughes-Croucher, John Keagy, Yousef Khalidi, Lew Tucker, Richard McAniff, and Brian Wilcove.

Thorsten: I bow my head to my long-time mentors and rock stars Holger Spielberg and Lars Bodenheimer. Your constant support, constructive feedback, and creative minds have always been a role model for me.

This book is dedicated to our partners, wives, families: Thank you for being patient with us when we had night shifts; when we were away on conferences and meetings; when we told you "just five more minutes"; when we were giving you the bottle while on the phone and typing emails; when we had frozen pizza for dinner – again.

I hope we got this right – you tell us in 15 years…

Contents

Introduction

The American economy had experienced tremendous growth on what I would call the first wave of Internet based businesses, and the question mark is to what extent we might see a second round of hyper-growth associated with the Web. We believe that Cloud Computing may offer that next wave of economic growth with businesses and startups flourishing alongside larger firms entering the cloud, and spurring a new wave of job creation and economic prosperity born out of the technology sector.

This is a nascent industry and there continues to be a need for R&D to build out some of the longer term capabilities that have to be put in place. For Government to procure cloud services in a meaningful way, we have to validate standards for interoperability: so that if we were to place our data in a cloud provider, how do we have an exit strategy to move that data, to avoid vendor lock-in. We also acknowledge that addressing security issues in the cloud is key to avoid the cataclysmic effect of having all of your data

Aneesh Chopra
Federal CTO of the United States

corrupted and unauthorized accessed. When we see IBM, Google, Microsoft and others all start to say they will move their enterprise class software into the cloud it opens up a much more interesting dialogue.

Our CIO, responsible for the $76 billion in IT budget management, is evaluating the right way to shift our resources from lesser performing assets to building new data centers, continuing our $19-plus billion of infrastructure spend, and migrating our federal agencies toward a more shared infrastructure environment. Whether we call that private cloud, public cloud, public-private cloud, there may be various sets of operational elements here that we as customer will help to shape the market for. So it all blends together: long term research, collaborating on standards, and then acting with our own resources so that we make the right investments to participate.

How can telecoms and other infrastructure providers learn from our initiatives and how can they make sure to make the right moves at the right time?

An article in Harvard Business Review by Jeffrey Immelt highlighted GE's approach in this regard. They call it *reverse innovation,* and it really speaks to the translation of C.K. Prahalad's vision into a corporate environment. Now why do I bring that up? GE has introduced medical diagnostic equipment at 85% of the price point, which is profitable

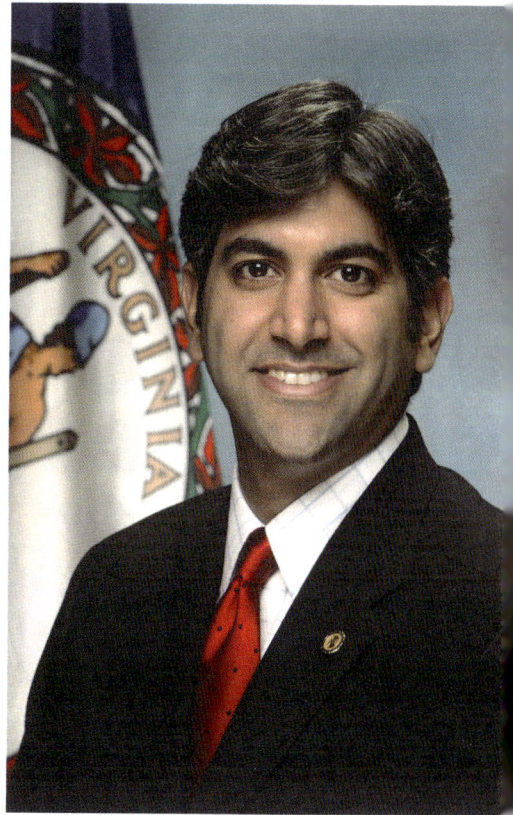

in rural China. We can re-import that technology and innovation into the US domestic market to expand the marketplace – redeployed to the US with more modular design, more sharable intellectual property. I see tremendous opportunities looking at emerging economies as target markets for US firms wishing to exploit principles of reverse innovation: Innovate in these countries where price constraints are such that traditional services are too costly. Gather that feedback and learning, and re-import that feedback as GE describes. I think that same play will be played out in the US technology sector. And if we don't play it, others will.

Cloud is an effective tool, platform, and business concept to drive reverse innovation. And that is why I think Cloud is an opportunity for revenue growth and an opportunity for the next wave of success in the technology sector that fosters job growth.

About This Book

Clouds. Immense, amorphous, and constantly evolving. With no definite shape and structure, clouds present a convenient way for us to represent something that is somehow inherently complex and yet simple. Something formless, without boundaries and yet highly structured. Something that is ever-changing, and yet eventually always the same. Something that makes representation of "stuff out there" remarkably simple in Microsoft PowerPoint. And you can create a whole new terminology and spin of your specific business or industry to Cloud. That explains why there are myriads of definitions and descriptions for the latest application of Cloud in the IT landscape.

This is not a book about Cloud computing, Cloud security, or Cloud whatsoever! I'm not going to discuss definitions of Cloud – again! I'm not going to pitch you any consulting practice or service offering. You probably were already browsing through the Amazon.com selection of roughly 400 books on "Cloud Computing", or through over 700 books on the German Amazon.de – ah, Germany, country of poets and writers. There you are going to find anything from "untold stories" (now finally told) to industry specific books in healthcare, government, retail.

I'm not saying that these books are silly or not useful – on the contrary: if you're a telecom, with a fixed-mobile converged service offering, operating a diverse array of networks, catering to consumers and businesses and enterprises across all industries, operating data centers and billing infrastructure, you probably have to read many of them. My personal library has only about twelve of them as hardcopies, my employer has another nine, and our knowledge management portal – internally known as MERLIN – has about 700 research reports from the usual suspects plus about 300 global project deliverables and documents we run in the past four years across all industries, including telecoms. And this is exactly the problem.

Beyond operation of communication networks, other core competencies of telecoms are vendor management, large-scale infrastructure management and refresh, and with it outstanding debt financing and management. I'm not downplaying telecoms' innovations and their often pioneering work, especially within tier two and three carriers. But their CTOs and CIOs will very likely agree that cost control and financing of their projects remain an everlasting battle, no matter how smart your employees and how cool your technologies are.

Because of these large infrastructure investments their infrastructure refresh rate is rather slow – especially compared with the accelerating digital services innovation cycles. Most of these infrastructure investments are multi-vendor ventures with innovative and often proprietary solutions to otherwise common architecture building blocks. There are two things I learned from my past projects in telecoms around Cloud: A) Telecoms will depend on their vendors' roadmap over the next five years and the evolving Cloud ecosystem they have to play in and influence; and B) the long term technology and service innovation roadmap over the next fifteen years – as any of their current infrastructure and architecture investments will be around for a while.

I couldn't find a book about that. I also didn't set out to write one about it, I'm not an expert. But I started to talk to the experts among us, who are not just working in the area of Cloud, but who are actually defining what the Cloud is and what it will be. These experts came from all kind of different industry backgrounds: telecom vendors, suppliers, customers, venture capital, government, and research and development. I quickly discovered that my fine contacts and conversations were far too precious to keep them to myself. I kept relaying their words and views, and a quick walk to the water cooler became hour-long strategy discussions. I also quickly realized that rather than trying to boil down the insights and learnings from these experts into comfortable, neat little packages of answers (or even worse – consulting frameworks!), it was far more appropriate to share their opinions and knowledge in their entirety. I'm a fan of Walter Murch, and "The Conversations" by Michael Ondaatje was one of the most insightful and inspiring pieces on film editing I've ever read. I'm also a fan of Feynman, and he has some good advice for us as well:

> If you don't like it go somewhere else! Maybe some other universe, where the means are simpler, philosophically more pleasing, more psychologically easy. I can't help it, ok?! If I'm going to tell you honestly what the world looks like to human beings who have struggled as hard as they can to understand it, I can only tell you how it looks like. I'm not going to make it any simpler; I'm not going to 'simplify' it, eh?! I'm not going to fake it.

And so we have the format of this book. We have included a short introduction that maps out the journey of Cloud to this point and presents some of our views and opinions on its evolution, but the bulk of the book concentrates on sharing the detailed views and immense expertise of some of today's thought leaders in the space, some of them more than nine months ago. To do justice to their insights, and to present their views fully, we have included all of the discussions in the actual format in which the responses were given. So the book consists of a number of interviews, each one previewed with a short abstract of the highlights and key take-aways from the discussion, to help guide you towards the most relevant topics for your interests.

The theme of this book, as the title suggests, is concentrated on the specific impact, challenges, and opportunities of the Cloud evolution for the telecoms industry. We all work for a telecoms telecom, so this is without doubt our industry and area of expertise. And so it's logical that the impact of Cloud on our industry is the focus of our efforts. But more than this, from all the literature that we've read, colleagues we've discussed with, and conferences we've attended, we truly believe that there is a significant gap in the understanding of the Cloud model as it applies for the telecommunications industry in the medium to long term future. To what extent do telecom telecoms need to utilize Cloud within their own operations? Will telecom telecoms have a role to play in the Future of Cloud? Or will it be the traditional Web providers – Amazon, Microsoft, Google, Salesforce – that will dominate? Will telecoms providers be able to play in the services and applications space? And how can the fast-paced, on-demand world of Cloud be reconciled with the telecoms' long-term, steady-as-she-goes value creation goal?

We are sincerely grateful to all of our interviewees, who were kind enough to share their time and wisdom. We believe that the combination of their opinions on the Future of Cloud makes for a tremendously interesting and relevant overview for telecoms today.

Ah, Clouds.

Thorsten Claus, July 2010

The Evolution of Cloud

How Did We Get Here?
The Next Five Years

business

business models

telecom

SalesCo

UI

level

telecom

external

units

relationships

management

customers

operational

software

business models

models

markets

Web 2.0

company

provisioning

reducing

innovation

new

create

ServCo

SOA

areas

3rd party

brand

platform

provider

success

service

resources

product

service provider

infrastructure

core

vendors

costs

differentiation

focus

way

specific

sourcing

business unbundling

sales

network

build

market

development

business business unbundling

servers

computing

horizontal

How Did We Get Here?

At its core, any business has three main business processes: customer relationship management, product innovation, and infrastructure management. Business Unbundling describes the effect when these three core business processes become separated by specialized companies due to reduced interaction costs within and between companies and customers. The operational model of Web 2.0 is helping to massively reduce interaction costs, subsequently increasing market fragmentation. In order to deal with the operational model of Web 2.0 and the associated fragmentation, the ICT industry required a design framework that allows flexibility while providing operational stability. The principles of Service Oriented Architectures (SOA) from the early 90ties flourished. If Web 2.0 is an operational model, and SOA its design framework, then Cloud is the (business) model for service provisioning and exposure to external and internal parties.

I'm not a historian. So to paraphrase Feynman: "What I am telling you is a sort of conventionalized myth-story that the consultants tell to their clients, and those clients tell to their clients, and is not necessarily related to the actual historical development, which I do not really know!"

Web 2.0

The dot-com era was built on the Internet. It was an era of firsts, of originality. We saw the first web browser, the first open source models, the first self-service portal, the first social network, the first digital photo sharing and printing services, and the first unified communication suites. Or at least a glimpse of what it could be. The important part – besides many lessons learned about failure and financial risk – was the rise of the operational model of Web 2.0. I call it an operational model because it has nothing to do with technologies.

While the term "Web 2.0" had notably been around since 1999, Tim O'Reilly and John Battelle consolidated its meaning to an aggregated form around the Web as a Platform[1] and a customer-centric business approach. In essence, they brought together these seven core elements of Web 2.0 (O'Reilly, 2005):

- The web as a platform
- Harnessing collective intelligence
- Data is the next Intel Inside
- End of the software release cycle
- Lightweight programming models
- Software above the level of a single device
- Rich user experiences

---

[1] The X-as-a-Platform and X-as-a-Service will haunt us for a long time to come, I promise you. Everyone is jumping on this bandwagon. Enscone Data Technologies even has a "Dead-as-a-Service", products that assure data on hard drives that is lost, stolen, or redeployed is destroyed beyond forensic reconstruction. Go figure.

Some companies were better than others in the execution of these operational elements[2]. These "best-in-class" Web 2.0 companies had some common core competencies:

- Providing services, not packaged software, with cost-effective scalability
- Controlling unique, hard-to-recreate data sources that get richer as more people use them
- Trusting users as co-developers and creating an environment or platform of trust
- Harnessing collective intelligence and creating an environment or platform of intellectual exchange
- Leveraging the long tail through customer self-service
- Creating and maintaining software above the level of a single device
- Combining lightweight user interfaces, development models, AND business models

Some of these core competencies are rather elusive: why should targeting niche markets make more revenue than catering to the masses, and what has the long tail to do with customer self-service? Why do I need lightweight user interfaces, why do my business models need to be lightweight, and what are "lightweight" business models anyway? The short answer to these actually more complex questions is: Core elements of Web 2.0 massively reduce interaction costs, something that is vital to survive in a world of business unbundling and re-bundling.

Business Unbundling

Interaction costs are money and time that are expended whenever people and companies exchange goods, services, or ideas. The exchanges can occur within companies, among companies, or between companies and customers. They can take many everyday forms, including management meetings, conferences, handset conversations, sales calls, reports, and memos.

In 1999 John Hagel and Marc Singer published an article in the Harvard Business Review called *Unbundling the Corporation* (Hagel & Singer, 1999). The article is focusing on the economic transformation of verticals across the sectors of the global economy – something telecoms should be familiar with by now. In a nutshell, Hagel and Singer found that all business processes of any enterprise fall into three core categories – they call them processes – and that reduced interaction costs leads to an unbundling of these three core categories:

- CUSTOMER RELATIONSHIP MANAGEMENT has to find customers and build relationships with them: draw people into branches or stores; assist customers and try to build personal relationships; respond to questions and complaints, process

---

[2] Fortunately John and Tim had the foresight to keep adjectives kind of broad so that we are still struggling in fully executing all aspects of Web 2.0: the core elements do not state how large *collective* is, or how local it should be; more screens don't automatically create a Web 3.0, and *rich user experience* is something that continuously evolves. We'll probably laugh about HTML5 and Flash 10 when we look back in 10 years or so, just as we giggle now about the first "rich" websites we saw in 1990 with a – lo and behold – animated GIF.

returns, or collecting customer information. Although these functions may belong to different organizational units, they have the common goal to attract and hold on to customers.

- PRODUCT INNOVATION has to conceive attractive new products and services and figure out how best to bring them to market: research new products and ensure that the company is capable of bringing them to market successfully; constantly search for interesting new products and effective ways to present them to customers.

- INFRASTRUCTURE MANAGEMENT has to build and manage facilities for high-volume, repetitive operational tasks such as logistics and storage, manufacturing, and communications; this includes building new branches, cell towers and central offices, maintaining data networks, and providing the back-office transactional services needed to process orders and present statements to customers.

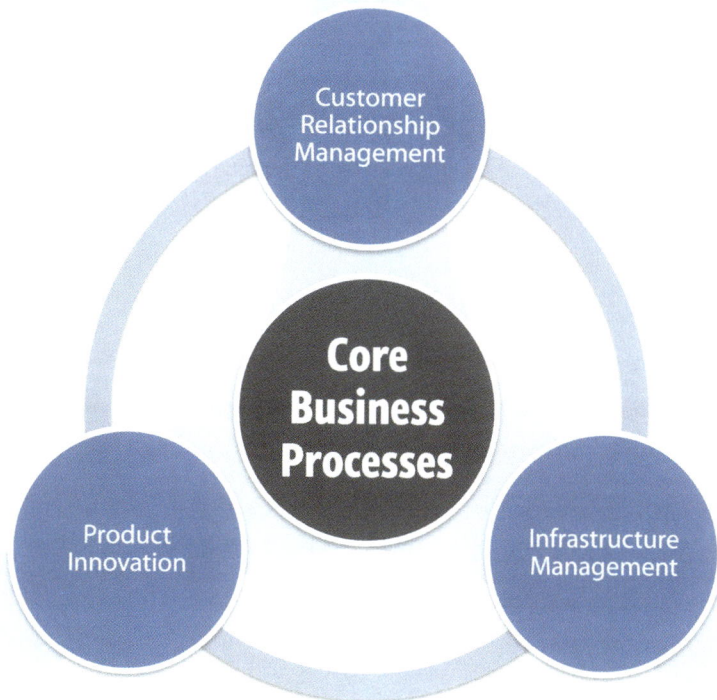

However, the business mechanics for these three core business processes are competing, and their divergent economic, cultural, and competition-related imperatives inevitably conflict and lead to compromises and trade-offs. A specialized competitor who does not have to make these compromises can offer specific elements of a core business much more effectively than a fully integrated company.

Telecoms know this all too well: other companies seem to innovate quicker, Facebook seems to be able to build a deeper emotional customer relationship on a larger scale, and so on. That wasn't an issue as long as interaction costs were quite high, as interaction costs could be used as market entrance barriers.

Unbundling of core business processes is actually also a chance to re-focus the business: an insurance company is very good at creating insurance products and assessing and managing its customer relationships – why should an insurance company run its own infrastructure? Well, in the past it had to, because there was no way you could "unbundle" that part of the business. A telecom operates a sophisticated network of infrastructure with complex business relationships and maintains a direct access into consumer wallets and minds – why should they focus product innovation? I'm not talking about abandoning new products and services, but about assimilating innovation, the art of integrating and managing new products and services into the portfolio landscape of the telecom. If infrastructure management is one of the core competencies of a telecom, surely systems integration and management could be a core competency as well?

It is precisely this observation that brings us back to Web 2.0: Web 2.0 accelerates business unbundling by reducing interaction costs. Its customer-centric approach, *"harnessing collective intelligence and creating an environment or platform of intellectual exchange"*, lead to app stores and open development initiatives. In a way, Web 2.0 opened the doors to cost effective out-sourcing.

And the secret to success in increasingly fragmented industries is not just to unbundle, but to unbundle and re-bundle, creating a new organization with the capabilities and size required to win.

Service Oriented Architecture

In order to prevail in the acceleration of business unbundling and market fragmentation even lower interaction costs are required, fueling the operational model of Web 2.0. Constant change of suppliers, services, vendors, and technologies is not something telecoms were very fond of so far, as a natural result of their focus on long-term strategic network investments. But they also couldn't stop Web 2.0, either. What they needed was a design framework. And not just any framework, but one that is

- Service and customer oriented;
- Allows the reuse of services, applications, and functional behavior;
- Permits agile change in business processes on top of existing services and information flows;
- Supports real-time monitoring of services and information flows;
- Allows the exposure of enterprise processes, services, and functions to third parties within and outside the enterprise.

Like so often such a framework already existed since the early nineties. This design framework was called the Service Oriented Architecture (SOA). With the need to keep step with the seemingly faster and hipper Web 2.0 crowd, every major telecom vendor jumped on the bandwagon and offered their spin on SOA. I am sure you have your very own flavor (or probably flavors) of SOA within your organization.

The Cloud Model of Service Provisioning

My guess would be that by now you've read countless articles, blogs, and books about Cloud and understand the general notion of Cloud, where you do some workload somewhere else, that "else" mostly being run on a shared infrastructure to minimize operational costs (and subsequently costs for you). You probably heard by now about the different deployment models of private clouds, community clouds, public clouds, and hybrid clouds. I could give you my own definition about these terms, but you will find that every vendor, book, and blog will have its own definition anyways, so I won't add to the clutter here – I think you get the general idea of Clouds being run in different domains internal or external to your own(ed) operations.

At this point you should have yelled "wait a minute – what you described so far is nothing more than the previous model of Application Service Providers (ASPs)! Throw some virtualization into your outsourced or hosted IT services, and you're running many different logical machines on a single physical machine off-premise. So why Cloud?"

Let's recap: You need to align your IT with a very specific Service Oriented Architecture to be compatible with the operational model of Web 2.0, while being able to source specific flexible and agile services from specialized external companies – 3rd parties – and to unbundle and re-bundle your business as needed. However, ASPs moved the complexity from your basement to their basement – you are now running the servers somewhere else, but they are still run on specific servers. You also usually only pay for the hosting, not the software itself, and ASPs also usually host one customer per instance of the software.

The complexity remains. What is missing is the business design of the operational aspects as well as architectural design: you probably need to do something different or new in order to engage and disengage quicker with 3rd parties and cater to the Web 2.0 operational model. These challenges along with some changed economics of datacenter and network operations created six characteristics for the Cloud-way to consume or offer services[3]:

- VIRTUALIZED RESOURCES: Resources are of virtual nature, separated from physical machines. Most likely, you actually share physical resources with others, creating the Cloud-typical multi-tenant environment.
- ELASTIC RESOURCE UTILIZATION: Resources can be added or removed as you need, only your application and service architecture is the limit. Adding blade servers are a physical example of elastic resources, spinning up more servers as you need more computing power.

---

[3] Of course it's a bit more complicated than that. But this is not a history class about Cloud Computing. I rather want to make the point that *Cloud* fits the need to populate an operational model within architectural design constraints with adequate business mechanics.

- **AUTOMATION:** You don't actually add or remove resources by hand; they rather get added or removed automatically, depending on the computing need at hand or some business rules. Again, this is nothing particularly new, you were probably spawning images of servers automatically before.

- **UTILITY PRICING:** While you're doing all this scaling and utilizing, you pay-as-you-go, or pay-as-you-grow. There are several different models here: paying by computing power or storage or bandwidth or volume, with and without a cap, a guaranteed free baseline with a premium once you reach a cap, etc. But the main idea is that you can start small and pay small – which, in return, might also end up the other way around: you end big and pay big.

- **SELF-SERVICE PROVISIONING:** You don't have to call a systems integrator, get a quote, wait for a week, and finally get all your services up and running. You probably have a web-based interface where you can do most of the activities yourself, or even have a way of automating provisioning.

- **MANAGED OPERATIONS:** The actual operations of the IT infrastructure required to provide necessary resources – hardware, software, connectivity, security, backup, etc. – is done by someone else whose core business it is to operate and manage this IT infrastructure. That might be an internal business unit of your company or an external 3rd party or a hybrid solution where you provide space and servers and connectivity that is maintained by a 3rd party.

As you can see, none of these six characteristics are particularly new or ground breaking. You are probably familiar with outsourcing, your IT department is probably already running virtual machines from different departments on the same blade server, and if you're a really large telecom you probably pay a utility fee for the IT services you use.

So Cloud is merely a new way of providing services. Probably the really new part about it is that infrastructure becomes a service. Cloud is, in a sense, a kind of business model. There are, in fact, many Cloud business models, each with its own value proposition, key resources and activities, revenue streams and costs, but I am alluding to the fact that Cloud is merely a new way of doing IT business, and not necessarily a new technology or service. At least not initially – but we will get to that.

By now you will have heard all the dreaded four-letter words: IaaS, PaaS, SaaS – Infrastructure-as-a-Service, Platform-as-a-Service, and Software-as-a-Service. You might have also heard some more functional descriptions such as Testing-as-a-Service, Management/Governance-as-a-Service, Application-as-a-Service, Process-as-a-Service, Information-as-a-Service, Database-as-a-Service, Storage-as-a-Service, Computing-as-a-Service, Security-as-a-Service, Integration-as-a-Service, and so on. (I told you the X-as-a-Service would haunt us!) These business and IT needs existed long before Cloud, of course, and were addressed before in one way or another. However, in the Cloud model of provisioning these services, the six characteristics mentioned above apply.

For telecoms, Cloud is a business model (again: in a broad sense) of providing services that caters to a Web 2.0 operational model and is manufactured within a Service Oriented Architecture[4].

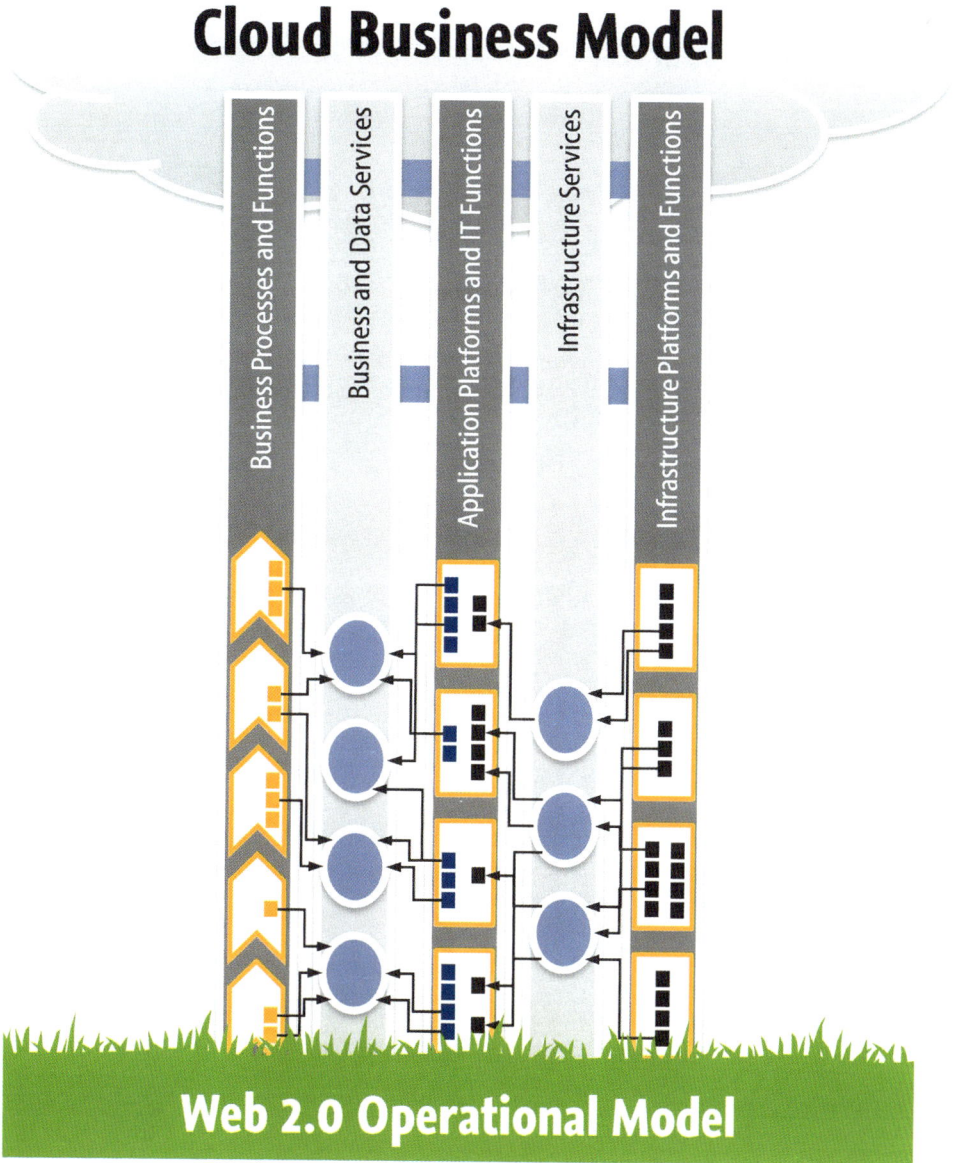

Cloud Business Model

(Columns, top to bottom: Business Processes and Functions; Business and Data Services; Application Platforms and IT Functions; Infrastructure Services; Infrastructure Platforms and Functions)

Web 2.0 Operational Model

---

[4] Of course it caters to a lot more and different models and architectures. In fact, exactly because Cloud is "just" a business model of providing services it is applicable to many different domains, industries, technologies, architectures, frameworks, operational models, etc. But as I am providing you with a "conventionalized myth-story" targeting telecoms, I will focus on standards, architectures, frameworks, and designs common at emerging and mature telecoms.

The Next Five Years

Current blogosphere hype, market segmentation, and service positioning only really addresses two of the eight focus areas of IT with sufficient consistency: A) Reducing costs and B) Reducing service provisioning times. To create a sustainable ecosystem of Cloud service providers who can address other focus areas of IT, segmentation and clustering into more defined markets and roles is necessary. The results are a consolidation into only two sophisticated markets – a market of infrastructure utilities and a software universe – and a value chain segmentation that will create three distinctive operational layers for Cloud service providers, the SalesCo, ServCo, and AssetCo. Current Cloud players are already positioning themselves accordingly.

The Role of IT Infrastructure and IT Strategy

You probably also read a lot about the benefits of Cloud. Depending on what aspect you're looking at and which industry you're working in there's a plethora of subjects that Cloud can help you fix: Cloud and social networks, Cloud and brand management, Cloud and hyperlocal content aggregation... you name it, Cloud has it! Once you dig deeper you also find out that most of the stated benefits are use-case specific and very context-sensitive. I'm not suggesting that the benefits do not exist or are made up, but it is often hard to put the finger on the concrete benefit and applicability for your specific organization. However, the most commonly agreed upon benefits for Cloud are

A) Reducing costs and

B) Reducing service provisioning times.

The second benefit is also referred to as reducing deployment time, reducing the time-to-market, increasing the speed to implementation, etc. This is great news! These two areas are definitely important fields of action for IT – internally as well as externally. But if you're the CIO or CTO or working in these business lines "cheap and fast" is not the only thing you need. Other areas of action and interest are just as important, such as:

C) Increased speed of innovation

D) Increased precision of capacity demand forecasting and planning

E) Reduced vendor and supplier dependencies

F) Support for business agility and enabling new business

G) Reduced interface complexity

H) Increased service quality

You might have noticed the A) through H) listing instead of 1) through 8) numbering. This is deliberate. I want to avoid the impression that there is any kind of common industry agreement on their prioritization. There isn't. In fact, when you look at academic research on strategic alignment and IT Strategy from Henderson and Venkatraman (Henderson & Venkatraman, 1993) or Raymond Papp (Coleman & Papp,

2006) you probably have very different perspectives that depend on how your IT is aligned and utilized to support your operations and business.

Development One: Divide into Two Sophisticated Markets

The diverse list of requirements will without doubt lead to a highly fragmented landscape. However, the way we introduce services and infrastructure will not change overnight, especially not within Fortune 500 companies with thousands of employees and – as is the case in Europe – a strong influence by regulators and unions. Furthermore, Cloud companies themselves will be under the same influence of unbundling as their customers. Cloud providers will also have to focus on one or two core business processes, operate on scale, or develop specialized service offerings[5]. A rift will develop on how to advertise, sell, and operate these services. And as a result, all X-as-a-Service offerings will have to play in one of two markets (and, as always, there are going to be exceptions to the rule).

SaaS
IaaS PaaS

Software Universe

Infrastructure Utility

- Industrialized IT infrastructure and middleware standardization
- Automated service provisioning, management, pricing, and billing
- Scale effects lead to price dumping and deteriorating margins
- Discounters and the competition on operational costs lead to consolidation of service providers, resellers of all sizes emerge
- Industrialized IT infrastructure providers compete on operational scale

- Differentiated and networked applications and frameworks across horizontals and verticals
- Modularized platform and application architectures
- Integration and orchestration platforms are emerging
- Sizes of service providers and players will range from one-man shows to large enterprises.

- **INFRASTRUCTURE UTILITY MARKET:** IT infrastructure is heavily industrialized and middleware standardized. Service provisioning is fully automated, as are management, pricing, charging, and billing. Scale effects will lead to price dumping and deteriorating margins. Discounters will eventually force a consolidation of service providers and vendors, and resellers will begin to emerge. While resellers will range from small businesses to large enterprises, industrialized IT infrastructure providers will compete on operational scale.

- **SOFTWARE UNIVERSE MARKET:** Applications and frameworks will be highly differentiated across horizontals as well as across verticals. They will build upon each other similar to the Web 2.0 concept of mashups. In effect, platform and

---

[5] Another way to see this unbundling is from the *disruptive innovation* point of view: providing new kind of services to unserved or under-served markets or providing better targeted services with less feature-overkill to existing markets, both of which could be accomplished on any kind of scale and organizational size.

application architectures will be modularized, also because of the pressure of decreasing interaction costs, business unbundling, and the operational model of Web 2.0. Integration and orchestration platforms are emerging to manage this unbundling, re-bundling, and agile service combination. Sizes of service providers and players will range from one-man shows to large enterprises.

Just to be clear: I'm not saying that there will be only room for five cloud computing providers on this planet. That would be pretty silly, as the regional regulations, operations, culture, skills, environments, and business models are quite different[6]. For example, there might be a point where the operational costs and the costs to insure operational risk for a computing cloud in twenty different countries is higher than the benefit of its scale. So if you don't need global operations, a smaller local provider without the feature overkill of cross-country governance and service assurance might be a cheaper way to go.

I'm also not saying that a company could not develop solutions for both markets, or that a solution could not be sold in both markets. Moreover, there is no preference regarding brand versus discount placement. Sometimes it might be a matter of branding and sales approach: Jimmy Choo and Nancy Gonzalez are selling their handbags at Neiman Marcus, Samsonite is manufacturing and distributing Timberland luggage and bags, and Deutsche Telekom is offering mobile pre-paid phones and DSL landlines under its Congstar brand.

Development Two: Unbundling of Business Models

The concept of Cloud became popular because of the demands and effects of business unbundling, as explained earlier. This effect applies to Cloud players and their business models as well. They also will have to focus on one or two core business processes as interaction costs decline. But as in any industry, no company would like to phrase their selection of focus as *"Oh, we decided to not do customer relationship management any longer"*, or *"Product innovation… really not our strength, we outsourced that…"* So Cloud players will find a new terminology – or actually a not-so-new one. The horizontalization of the industry, already well known to telecoms providers, defines three layers:

- ASSET COMPANIES, or AssetCos, provide basic computing and connectivity facilities. Datacenter operations of an IT telecom and product development of a vendor are examples for AssetCos. They have to be the best in cost-effective IT management and quality control. Their critical success factors are efficiently scaling facilities,

---

[6] Some blogs are quick to suggest that similar to UPS, DHL, TNT, EMS, and FedEx there will be only a few providers who will be able to operate on a global scale. But these companies are only so much in our minds when it comes to delivering and shipping goods globally because of their large consumer targeted advertising spend. There are many other global delivery network providers targeting specific verticals that are using none of the above five providers.

tight cost and overhead control, complexity reduction, and constant adjustment of their business models to new demand patterns and business models[7].

- **SERVICE COMPANIES**, or ServCos, provide services based on utilities operated and managed by AssetCos. Operational and integration services of IT telecoms and support and professional services of vendors are examples for ServCos. They have to be best in innovation. Their critical success factors are relentless product, solution, and service innovation, agile sourcing of utilities for best service experience, perfect market timing, and a rapid time-to-market.

- **SALES COMPANIES**, or SalesCos, are the customer interface to services provided and managed by ServCos. Account management of IT telecoms and distribution and sales channels of vendors are examples for SalesCos. They have to be best in customer intimacy and experience. Their critical success factors are efficient sales and point-of-sales management, customer service, brand value, and agile sourcing and discovery of services from ServCos.

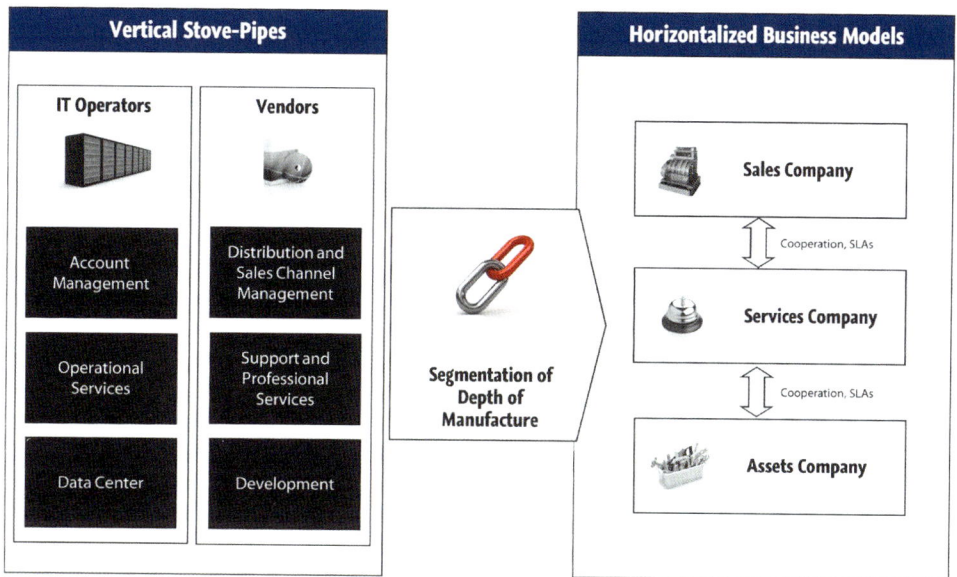

Vertical Stove-Pipes		Horizontalized Business Models
IT Operators	Vendors	
		Sales Company
		Cooperation, SLAs
Account Management	Distribution and Sales Channel Management	Services Company
		Cooperation, SLAs
Operational Services	Support and Professional Services	Segmentation of Depth of Manufacture
Data Center	Development	Assets Company

Horizontalization will decouple the different business layers within a Cloud service provider. Horizontalization also has the connotation of breaking vertical silos, now serving more customers in many industries. Business agreements and contracts between these layers will include Service Level Agreements (SLAs), providing the basis for collaboration between these three layers.

New Cloud service providers will try to focus their operations on one of the three specific layers within one of the two different markets from the beginning. Already

[7] The business model agility is often neglected, but a crucial element for AssetCos: You might provide best-in-class utility computing, but if your revenue model is, for example, based on billing per transaction and all state-of-the-art services constantly pumping micro data back and forth, your business model is probably not going to be compatible with the cost it puts on your client's services.

existing Cloud service providers will start to horizontalize their business, first internally, providing all three layers of their existing operations through in-house provisioning mechanisms of dedicated business units, with defined SLAs between them. In the beginning the decoupling will allow the Cloud service provider to offer specific services into other organization that lack this specific horizontal core competency. A horizontal business line becomes multi-tenant enabled for internal or external consumption. Once the ecosystem has evolved and companies have established themselves as viable solution providers within one of the three layers and one of the two markets, the horizontalized business units of the Cloud service provider will allow a more cost effective and agile (out)sourcing to specialized third parties. The horizontal business layer becomes one tenant within a third party's multi-tenant Cloud offering.

Two effects of horizontalization that we can already observe in the world of telecoms are:

- The high level of standardization in AssetCos versus the comparatively high level of customization in SalesCos; and
- The small number of players at AssetCos versus the comparatively large number of players at SalesCos.

Why Do I Think That's Going to Happen?

So why do I think these alignments will happen – now or within the next five years? Because the current hype and positioning only really addresses two of the eight focus areas of IT management with sufficient consistency: A) Reducing costs and B) Reducing service provisioning times. One could argue that basically all measurements in IT are geared towards reducing costs and increasing efficiency. But my point is that there is an inherent opportunity for service provider differentiation if the other six areas of IT management are not sufficiently addressed. Once all areas can be addressed by an ecosystem of technologies, services, and business models – beyond just "cheap and fast" – then Cloud becomes a serious way of doing business.

But keep in mind that this opportunity for differentiation is there for any service provider, not just Cloud service providers. And other service providers already have critical mass and market attention. Collective marketing efforts that can harness collective intelligence[8] are required to create sufficient gravitation for a sustainable ecosystem of Cloud service providers – a consolidation of the current markets, trends, hypes, products, services, and technologies that are all necessary to multiply each player's effort. At the same time, all core characteristics of the operational model of Web 2.0 as well as business unbundling and re-bundling still apply, and are ongoing.

The two developments I discussed create a valid segmentation with Cloud business model categories that exactly address the six other focus areas of IT:

---

[8] Remember? We were talking about the operational model of Web 2.0, core element two: "harnessing collective intelligence"! That operational mode is still active and working, also for Cloud service providers.

- In INFRASTRUCTURE UTILITY

  - AssetCos will increase the service quality through Infrastructure Factory business models, providing datacenter and network operations, infrastructure lifecycle management, and hardware manufacturing and maintenance.

  - ServiceCos will reduce vendor and supplier dependencies through Lean Service Provider business models, providing infrastructure aggregation and value-added operations.

  - SalesCos will increase the precision of capacity demand forecasting and planning through Cloud Resource Broker business models, providing wholesale and retail channels and a cloud resource exchange.

- In SOFTWARE UNIVERSE

  - AssetCos will increase the speed of innovation through Software Development business models, providing cloud-aware applications and middleware or platform development.

  - ServiceCos will reduce interface complexity through Application Integrator business models, providing integration platforms and network identity repositories.

  - SalesCos will support business agility and enable new business through Cloud Service Distributor business models, providing horizontal app stores and vertical market sales channels.

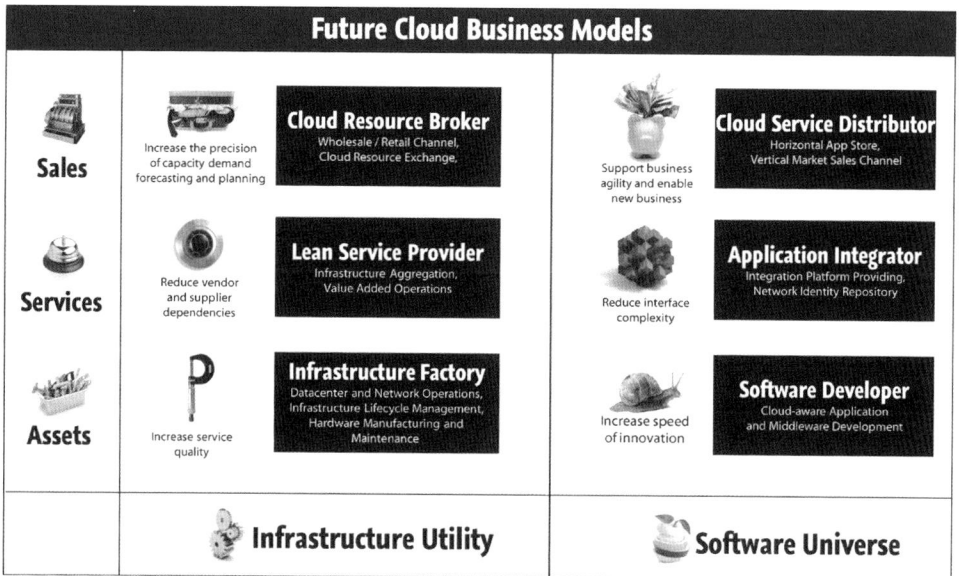

Future Cloud Business Models

Sales	Increase the precision of capacity demand forecasting and planning	**Cloud Resource Broker** Wholesale / Retail Channel, Cloud Resource Exchange,	Support business agility and enable new business	**Cloud Service Distributor** Horizontal App Store, Vertical Market Sales Channel	
Services	Reduce vendor and supplier dependencies	**Lean Service Provider** Infrastructure Aggregation, Value Added Operations	Reduce interface complexity	**Application Integrator** Integration Platform Providing, Network Identity Repository	
Assets	Increase service quality	**Infrastructure Factory** Datacenter and Network Operations, Infrastructure Lifecycle Management, Hardware Manufacturing and Maintenance	Increase speed of innovation	**Software Developer** Cloud-aware Application and Middleware Development	
	Infrastructure Utility		**Software Universe**		

You could argue that anyone could come up with some sort of two by three matrix shown on the next page, throw in the six not-so-well addressed focus areas of IT, stir (don't shake), and voila! a consultant's wet dream of pointy boxes is born (you got me – that's how I did it!). However I might have come to my conclusions, it is certainly apparent that the current Cloud service providers are already starting to position themselves accordingly:

- In the market of **INFRASTRUCTURE UTILITIES**,
  - Rackspace, GoGrid, or Amazon are Infrastructure Factories,
  - Scalr or RightScale are Lean Service Providers, and
  - USourceIT or Ayuda Networks are Cloud Resource Brokers.

- In the market of **SOFTWARE UNIVERSE**,
  - Salesforce, Taleo, SpringCM or HyperOffice are Software Developers,
  - Cast Iron Systems, Boomi, or T-Systems are Application Integrators, and
  - Deutsche Post DHL, Google with its Apps Marketplace, or Apple are Cloud Service Distributors.

There are other hints as well. Earlier I explained how existing Cloud service providers will first horizontalize their operations to be ready for unbundling and re-bundling their business once the ecosystem has enough mass to allow agile (out)sourcing. This process involves a reorganization of business units, organizational structures, and people. And that is exactly what is happening right now:

- We see a larger than usual fluctuation of people between Cloud players.
- There are a larger than usual number of job openings with more specific position and job descriptions than just Software Architect or Marketing and Sales Executive, targeting specifically either the infrastructure utility market or the software universe.
- You can search for recent hires at Cloud players in business social networks like LinkedIn and Xing, to see that there is a high fluctuation of senior leadership between Cloud players. Actually, while we were preparing this book, Raju Gulabani switched jobs from Google to Amazon, Lew Tucker away from Sun Microsystems to Cisco – and they are all heading into a bright future.

The Future of Cloud

What's Next?
Strategic Impact on Mature and Emerging Telecoms

devices
small flexible users content systems
stuff problem experts architecture security Internet step
patterns NetService client power
large solutions standards
people integration operational picture start
Cisco things processes operations
answer
access level cash
customers complexity mobile firms
examples future place need industry order number relationships
maybe developers tool explains Salesforce month acceleration
everywhere Microsoft companies computing platform footprint
capital course market
facebook probably else many storage real-time
price dynamically Brian use
networks space innovation
U.S compute managing
multi-vendors New business today connection
looking ideas device monitoring Allen utility
interesting towards delivery change CIO cost contracts
changes fact interoperability time Google advantage CTO lead
challenges
services call
federal
Jamie cloud
want communication role value chain Keag Successful
check Twitter different question
new bill walled garden aware
support data vendors experience secure provider
models connectivity resources businesses
opportunity communications providers might big phones way
applications move
mistakes effect

What's Next?

Telecoms will have to source and provide IT from serverless boutique integration firms. The new flexibility in sourcing and provisioning will also require more flexible cash flow configurations. Customer service will diffuse into the Cloud as well. Successful telecoms will find an operational mode for the strange combination of IT, network management, and marketing in order to get in front of their accidents and issues as they happen. Mobile devices will change communication paradigms and require a new kind of Cloud-based management and application acceleration. While secure, reliable, and often low-latency broadband access will be crucial for successful Cloud services, IT communications has been resolved to one commodity: IP transit. Telecoms will have to refocus their business and evolve from service providers to service brands. But there is no silver bullet for telecoms' positioning in the value chain – anything from low-end utility provider to high-level data analysis provider is possible. But telecoms will have to partner and cooperate to compete with global players like Microsoft, Google, or Amazon. Governments will accelerate the search for interoperability.

Telecoms are used to calculate business plans for their infrastructure investments over 15 years. Now we just have to look at the five month half life of a mobile device and we know that infrastructure investments are changing, even more so as telecoms move into services and applications.

I asked most of my interviewees the question what we would talk about in ten or fifteen years from now on. Unfortunately they were all smart enough to gracefully divert this silly question. Ismael Ghalimi brought up an interesting comparison on his first encounter with the NCSA (National Center for Supercomputing Applications) Mosaic web browser in 1994, and how his answer about the future of IT fifteen years from then – 2009 – would have been very different just a month before that encounter.

What my interviewees fortunately did share, however, were first recurring patterns and ideas that they are seeing within their specific industry or business, patterns and ideas that seem to stick. While they were quick to add that they have no clue whether these patterns and ideas become trends to influence a whole industry or whether they are just flashes in a pan, together they formed a scarily coherent picture of the future of Cloud in five to ten years. Of course "only Forms possess the highest and most fundamental kind of reality, and not the material world of change known to us through sensation," as Plato would say. But let's go watch some shadows.

New Ways of Doing Business

"There's a new class of boutique integration firms, companies that are building completely serverless IT portfolios for companies." says Peter Coffee from Salesforce. In effect systems integration arms of Telecoms will increasingly compete with boutique integration firms with much lower initial capital spending and more flexible service contracts. On the other hand, these companies will require reliable and secure operations. With the right mix of co-opetition Telecoms will be able to grow their Cloud services business if they are able to let go of some of their traditional end customers. Luckily Telecoms can forge strong alliances with their often already existing partners. *"Microsoft always believes in partners, and collaborating with partners in creating a big ecosystem for the benefit of everybody. We will not pursue this as a stand-*

alone thing." Explains Yousef Khalidi from Microsoft. *"Integrators will have to invest more and more in Cloud practices that are more meaningful and more targeted at the Cloud. I do believe that customers will want to deal with trusted companies that they know and have strong relationships with."*

A large stumbling block for telecoms will be the need for flexible reconfiguration of cash flows. Telecoms in the past were excellent in tightly managing their procurement process and negotiating longer term contracts with tough terms and low margins for vendors and suppliers. Telecoms thrive on gate processes, offer boards, and prioritization boards. *"[But] if the essence of Cloud is agile service re-configuration and sourcing [then] not only SLAs or protocols or APIs need to be reconfigured, but the flow of cash will have to be fast and flexible. Cloud players will integrate payment solutions to facilitate faster transactions. M2M authorization of cash flows – sounds a bit like 1999 Dot-com all over again."* says Peter Coffee from Salesforce.

Telecoms will also have to take a fresh look at the level of ICT innovation in industries. Richard McAniff from VMWare explains: *"Often, we forget that a lot of the applications we run internally that make up organizations and enterprises are already cloud. They're pulling out their CRM or HR system or financial systems and so on, and moving them to the cloud. So are they the first adopters? They may not think that they are adopters but they are already using cloud applications."*

Another challenge for telecoms will be internally to decentralize purchasing and decision power, and externally to commit to and sell service contracts to sub-divisions of their clients that were traditionally complex communication and IT services authorized and signed by the CIO or CTO. *"It's the tenant [who pays]."* says Ismael Ghalimi from Intalio. *"Let's take the example of Mitsubishi Corporation. Mitsubishi Corporation is the largest company in Japan. Their revenue amounts to six percent of Japan's GDP, and they're 400,000 employees, 1,500 subsidiaries. To answer your question about who pays us: it's the subsidiaries, and typically the department within the subsidiary."*

As more services migrate to the Cloud, a higher level of cross-company communication, co-operations, and know-how exchange will start. Former competitors will share common utility systems and functions because the joint development, maintenance, and upgrades will be cheaper than monolithic customized systems. This development will also move help and support out of the control of telecoms. *"[Telecoms need to] re-conceive the fundamental mission of a service and support tool given the existence of the Cloud."* says Peter Coffee from Salesforce. *"Because customers today no longer pick up the phone and call your toll free number and call you as step one of a problem-resolution process. The first thing they do is go online, check Google, check their community on Facebook, check Twitter, check any number of other resources where they think they can find a credible, independent expert solution and maybe, as the last resort, they call you."*

Because of increasing real-time data integration and flexible IT system and business relationships, liability becomes an issue. *"If web site content becomes dynamically generated from many sources and application spin up and down by themselves and these are 3rd party apps, who is liable?"* asks Pete Grillo from Iterasi. *"There are frivolous lawsuits that major corporations get two or three or five or ten times a week where your website was supposed to have said this. They want to know for sure when it said that."* Successful telecoms will specifically focus on two things: First, every IT project and system will include a monitoring and reporting mechanism, and every mission critical system will have alternative vendors for rapid fall-back and systems switch. Second, reporting and monitoring will be tightly coupled with corporate marketing and communications. This will be a more challenging configuration and way of thinking for telecoms. Telecoms need to get in front of their mistakes, outages, and system changes immediately if they want to maintain the high level of trust by their customers. Customers will have fall-back solutions themselves, just like telecoms, and while accidents and mistakes happen, the cost for switching telecom providers will be much lower than before. Telecoms need to pro-actively communicate and mitigate issues and outages with creative customer service on all communication channels their customer use. That will include Twitter, Facebook, and whatever next generation real-time communication tool there will be in the future. A press statement or a note on the telecom's website will not be sufficient anymore.

Telecoms will also have to utilize Cloud services and service provisioning to experiment and innovate outside their network footprint in order to accelerate innovation and lessons-learned and to subsequently re-import these innovations. U.S. federal CTO Aneesh Chopra explains: *"GE has introduced medical diagnostic equipment at 85% of the price point, which is profitable in rural China. If we re-imported that technology and innovation into the US domestic market to expand the marketplace – redeployed to the US with more modular design, more sharable intellectual property. So I actually see tremendous opportunities looking at emerging economies as target markets for US firms wishing to exploit principles of reverse innovation. Innovate in these countries where price constraints are such that traditional services are too costly."*

New Role of Devices

The explosion of mobile and personal devices will fuel Cloud services. *"Your cell phone, your smart phone with the Cloud, is the answer to the future of computing. Cell phones are dropped in the water all the time. Having any kind of permanency on a cell phone is ridiculously crazy."* says Pete Grillo of Iterasi. Of course the limited computing and storage abilities compared to the rapidly increasing data generated by end users and systems will create new challenges, especially for access to documents and media. *"What I see more and more is: People will expect an experience, beyond what is possible with a single machine or device."* Says Tom Hughes-Croucher from Yahoo. Timo Bauer, NewBay, agrees: *"Today most people think about how to get content onto certain devices. The 'sync' paradigm is all over the place. We think that the 'sync' paradigm is actually 'anti the cloud'. It is not about 'syncing' content rather about enabling easy 'access' to content that sits in the cloud."*

The network itself will have to become much more intelligent. *"If it's a picture that I have looked at recently it should stay on the phone, there's a probability that I'm probably going to want to show it to somebody else or look at it again. But if it's a picture that is three months old and I haven't looked at it, and it's already been uploaded to the Cloud – get rid of it. And then, if I do download it from the cloud again, don't send the four megabyte file down to my phone, I only have a little teeny screen. Transcode it and send a little teeny picture down."* says Jamie Allen from NetService Ventures.

Backward compatibility of devices will also become an issue. Telecoms will have to figure out a way to manage not only a large number of new devices, but also a large number of different communication paradigms. *"The initial architecture of all the mobile devices when they came was designed for client-server kind of applications. When you come in with a cloud computing architecture like we have with Google Apps, you have to take these devices and even though they believe they're talking to the old-generation architecture, they need to talk to the new generation cloud computing."* says Raju Gulabani from Google.

But the price point of thin client computing devices paired with the power of the Cloud might also have an altogether different effect on how we use devices around us. Ismael Ghalimi explains: *"The devices will be completely irrelevant, so today I see that most people have a lot of data on their laptop or their cell. In the future I have none. In the future I don't really own devices, 15 years down the road. Interfaces surround me everywhere that allow me to do stuff, to be productive wherever I am. So there will be computers, essentially on public access. They're just terminals. They're just dumb, and everything is on a server somewhere. And maybe I'm still carrying a device. Most likely I'm still carrying a device that we call a phone, the mobile phone today, but it's just one of these very, very many devices that I'm going to get."*

New Ways of Running and Providing Services

Successful Cloud services will be very much aware of their context. *"One area where I expect to see some radical changes is how the use of dynamically created virtual machines will cause us to change the way we design applications. With this new model, we will start to built applications as systems that are in some sense, aware of their own resource utilization, and can take steps to either increase the number of servers in response to increased demand, or decrease their resources to save money."* says Sun Microsystem's Lew Tucker. Peter Coffee from Salesforce agrees and extends an application's awareness beyond its own costs to the cost on shared resources: *"[You need] application-aware code: So you can write applications that provide a well-behaved user experience while at the same time respecting the need to avoid unreasonable burdens on shared resources."*

Many experts see a telecom's detailed knowledge about their end customers as an area of strength to provide customized or personalized services: *"Optimally, what you want is a very targeted set of applications, pricing, network access – for you."* says Sofinnova's Brian Wilcove. *"And in order to get that you need a bit of virtualization technology in order to sub-divide or cut customers up in a different manner so that you can bundle solutions for that individual. I may not be a Facebook user but I may want ESPN and some other stuff. And the telecom is a perfect channel, perfect conduit to deliver me those services. Another good example of, which I think is probably less than 15 years away – maybe five – is personalized TV."*

But telecom personalization and services will get strong competition from local service providers. *"The school system feels very comfortable working with a local partner and they are building an application to manage their schools more effectively."* says Richard McAniff from VMWare. *"[Local service providers] are aligning in a very interesting way with developers who are providing applications with VMware running the cloud, accommodating small business in their city."*

Because telecoms are good at managing multi-vendor environments in their own network, Aruba Network's Paul Curto is predicting that telecoms will leverage their own experience and know-how to provide the same multi-vendor managed services for other companies: *"We could see, for example, service providers getting into this space because if you're going after especially small and medium businesses and you don't have a large sales force, Management-in-the-Cloud is one way to set up the service, market it, and push it out towards these types of businesses, maybe work through a value-added reseller type channel model and develop your business that way."*

Paul also sees an opportunity for application acceleration: *"When you look at some of the other things, some of the other challenges to tackle, one of the areas to watch is application acceleration. When you think about it in terms of content delivery there are some challenges around content delivery networks not being that enterprise-grade or - secure. You will see a combination of enterprise-grade security with content delivery with end-to-end encryption. You could leverage Cloud-based services, but you would have to have kind of a hosted model in order to have support for that."*

Application acceleration specifically plays a role for the many different form factors of devices a telecom will see on its network. Differentiated application playout could lead to significant cost savings in network provisioning and vastly improved customer experience, argues Cisco's Simon Aspinal: *"Networks and the datacenter need to be aware that the user has a device with a small touch screen, so I can repurpose the content down to that device and size saving space, and because it is premium content therefore I need to prioritize it, over and above the rest of network traffic, this is the only way the user will get a good experience. It's that combination of being able to ensure a user experience while not relying on a best efforts networks, which would require enormous over-capacity to be built everywhere to be able to support that."*

Salesforce and VMWare both see a shift of programming and systems complexity away from the developer towards a platform. *"Application developers don't have to worry about the [complexity and reliability] system. You are going to reduce that complexity for the application developers. And it makes it easy for application developers to not have to worry about these kinds of complexities – they are not system guys."* says Richard McAniff from VMWare. But GoGrid's John Keagy is skeptical: *"[The most astonishing things in Cloud to me are reports of] features people supposedly want – what's supposedly important to customers and what's not. It's very different than what people like to blog about. There's a lengthy list of small items and some big architectural ones – hip and cool is often not addressing CIO and CTO requirements."* The problem arises out of the different views and needs between the CIO and CTO level and operational sub division. While in theory the goals and objectives of CTOs and CIOs are broken down into individual goals and objectives of division, the more operational and less strategic focus might prefer solutions that make programming easier, but increase operational complexity. This is further complicated because of the increasing divisional buying power for Cloud services. CTOs and CIOs will have to ensure that operational advantages for single divisions do not lead to operational disadvantages on a company-wide level. Successful telecoms will listen to their vendors' experience and best-practices across many different telecoms and other industries and pre-select Cloud solutions and services to avoid the wild growth of many monolithic sub-division-specific solutions that might only be of tactical advantage but of strategic disadvantage.

New Role of Telecoms

With data and services in the Cloud, connectivity becomes a problem. *"People will want to be sure that they have a high-speed connection to their Cloud data and that this connection and Cloud data is available all the time. The three things are speed, availability, and reliability."* says Raju Gulabani from Google. *"High speed connections become crucial, also for us at Google. We have 49 offices worldwide, and there is a culture of using video conferencing a lot, but that gets to be very expensive, getting conference rooms with hardware and so on and having all these connections. If you could do that kind of thing on a regular broadband Internet connection – that would be huge."* But this need for reliable, secure, and often low latency connectivity comes on top of the current vortex of HDTV and 3D TV. Telecoms are rightly concerned that their customers are not going to pay for the bandwidth. *"Obviously heavy smartphone data services are a major challenge for the carrier network infrastructure. One of the biggest issues for the carriers is that they have limited control over how applications interact with the Internet. The constant polling of social networking updates not only causes grief for network capacity planning but also results in a poor user experience due to decreasing battery life. There is plenty of room to optimize data services and I believe the carrier needs to step up and guide the community."* Says Timo Bauer from NewBay.

Beyond trying to get capital and operational expenditures for their networks down, telecoms need to look for other sources of income, playing on two-sided business models and revenue structures. *"[With the plethora of data running between client locations] it becomes the delivery networks that can track that stuff, for ad monitoring, for example."* says Iterasi's Pete Grillo. U.S. federal CTO Aneesh Chopra agrees: *"For the industry as a whole, analytics in the cloud I think is a tremendous opportunity, because more and more of us are gathering data at a rapid clip. The cell phone I have in my pocket emits location data every minute or so to the wireless towers. If you wanted a digital footprint of everywhere I have been you could literally look at the terabytes of data that telecoms might have on our patterns. They could then proceed to analyze that data and think about new and creative ways to monetize, without compromising privacy.'*

Initially telecoms will not have to throw their existing walled garden approach over board. There will actually be an opportunity in starting out as a walled garden, as Jamie Allen from NetService Ventures explains: *"There is an opportunity for the telecoms to provide a cloud based an application platform for their walled garden – good role model would be [NTT DoCoMo's i-mode] – an application platform that would allow application developers to deliver services to the telecom's broadband and mobile customers. A platform with integrated billing, customer support, and high security for people that are security conscious. The telecom can say, 'Hey, use this service. The data never leaves our network. It never goes out on the Internet'. There is a market for that, and nobody else in the world can do it."*

Iterasi's Ismael Ghalimi predicts that telecoms will have to become service brands, not service providers. *"The cable is not the vehicle for business, the billing system is the vehicle for business, the fact that adding a service to that bill is very easy and removing the service from that bill is very difficult. You've got a captive audience, captive customer relationships. It's customer ownership. People are going to identify themselves as a T-Mobile customer or an AT&T customer and will get their services, their applications. And the device doesn't matter, nor the service provider."*

But all these changes will eventually cannibalize existing services. Telecoms clinging to their old way of operations just because they conveniently fit with their current processes and target systems of their sales force will go under. *"I expect that over the next three years company phone systems will move to the Cloud at large scale, and I think it will start with small and medium businesses. There is no reason for a phone system to sit on site at all, besides some benefit to the liability."* says Google's Raju Gulabani. John Keagy from GoGrid goes even one step further: *"IT communications has been resolved to one commodity: IP transit. Telecoms have to decide if they want to be in the business of compute and storage. Network and access has been commoditized. Well, network costs are still making up a large part of the IT economy, but that's shrinking quite rapidly. Cloud compute and storage is so much analogous to networks because it's a scalable business that can be delivered on an automated basis where there can be some efficiencies had from scale."*

I don't think that there is a set place for telecoms where in the value chain they should position themselves. Some experts see telecoms at the low end of the value chain as a utility provider. *"A network is a Cloud."* says Brian Wilcove from Sofinnova. *"There's not much difference to me: the Cloud these days is basically the fact that it's more application centric, storage centric. It's the fact that applications have web calls as the interface to one another so you can mix data. Whether a telecom can become the next kind of step function in the economic value chain – I think that's probably true. Certainly, I think that's probably the place to attack the marketing. I'd go after the super low end."* Lew Tucker from Sun Microsystems sees telecoms advantage in the support and services space: *"Of course customer care has always been important. But the real-time aspect clearly shows the intersection of automation, self-service portals, and real-person interaction. Again, I think this is something telecoms actually understand. Support and service are going to be some of the real differentiators."* Simon Aspinal from Cisco stresses telecoms' excellence in vendor management, aggregation, and helping their clients to simplify service access and enable discover of the right services: *"I would love to see a significant change in the way services and interfaces could evolve beyond those offered today. There are millions of interesting Internet Cloud services out there. It's very hard to find them, and it's very hard to use them, and we still need that bridge to be able to get to them: a real simple interface and a simple and convenient way of finding them, I think that's a big opportunity for the telecom industry as this is classically what they do – engineer, market and simplify service complexity for the end user and cover the operational and bold challenges that go with that."*

Aggregation and discovery is not only a network and enterprise Cloud topic: *"Consider this, today millions of people share content across many sites without even reading and knowing the [terms and conditions]. The Facebook privacy discussion is just the beginning in this regard. So why not provide a technology platform to carriers that allows them to share all the user generated data with third parties, BUT also allows them technically as well as policy driven to retrieve the data back. The carrier will act as your trusted partner in this case and I believe no one in the value chain is better suited to do this better than the carrier."* Says Timo Bauer, NewBay.

But telecoms will also face tough competition from the usual suspects. *"It's hard to say which way it's going to shake out. The three major players in this space are going to be Google, Amazon, and Microsoft. Each of them is going to have strength and weaknesses within its space. I don't know what's going to be the winning formula right now."* says Google's Raju Gulabani. The problem becomes more severe as telecoms usually don't have the same global customer footprint and operations as a Google or Microsoft, and thus competing on low-margin utility computing might be a tough game to play in. However, telecoms have the advantage of many vendor relationships they could tap into. Cisco's Simon Aspinal explains how telecoms can success by leveraging their vendors' and suppliers' vested interest in global operations across carriers: *"We're seeing a very heavy shift towards on-demand consumption, initiated through web interfaces directly by a customer: Do-It-Yourself style models; pay by the day, by the minute, by the month, by the capacity, or by the compute power. So increasingly telecoms have to adapt to this more short-term, on-demand model which is leading them to the Cloud and the virtualization space. And we see a similar change in the way telecoms are looking to build that network: they often are looking more for assistance with build, operate, manage and transfer models; and the ability to be able to turn on features and capacities as users demand them."*

Tom Hughes-Croucher from Yahoo concludes: *"We need to get a better understanding and cooperation on many levels of the service production, end-to-end, through the Cloud. So that neither your, nor my brand is harmed, neither telecoms' nor Yahoo's service and revenues get impaired."*

New Standards

All interview partners agreed that standards would massively increase Cloud usability, especially for more flexible sourcing models and hybrid clouds. *"Vendors compete on the implementations, the excellence of system architecture, and the other services. I believe this leaves plenty of opportunity for the various players to compete and differentiate their offerings. Agreeing upon a standard API will simply help the market expand more rapidly."* Says Lew Tucker of Sun Microsystems.

But Tom Hughes-Croucher from Yahoo sees a bigger problem than consolidating or "harmonizing" a fragmented API ecosystem with rapid refresh rates: *"The API issue is frustrating, but you can code your way out of it. You can't code your way out of getting your data out. That is a network problem. I actually think that cloud peering will create a market and reduce some of the risks for cloud clients."*

While some industry standardization bodies started working on interoperability issues, the largest monetary commitment comes from governments who have a vested interest in non-proprietary data models that could lead to vendor lock-ins. The U.S. has the most prominent engagement and acts as a role model for many other countries. *"How can you ensure user authentication, identity, call it whatever you what, so you avoid the cataclysmic effect of having all of your data corrupted and accessed? So on standards for user authentication, and on standards for interoperability, we intend to engage collaboratively with the private sector on ways in which we can move these agendas forward."* explains an excited U.S. federal CTO Aneesh Chopra. *"We provided an additional $70 million to think about standards opportunities … what are the standards for interoperability: so that if we were to place our data in a cloud provider, how do we have an exit strategy to move that data, to avoid vendor lock-in."*

Strategic Impact on Mature and Emerging Telecoms

> *The world can no longer be divided into mature and emerging countries. Telecoms have to adjust their strategies accordingly. A Technology Metabolism index – the adoption rate of technology compared to a country's productivity – opens up a new way of thinking which countries might be quicker in adopting Cloud services than others. Telecoms should no longer see themselves as "emerging" or "mature", but rather define themselves by the environment they operate in. Telecoms in regions with higher technology metabolism will focus on becoming a Cloud Resource Broker, Cloud Distributor, or Software Developer; telecoms in regions with lower technology metabolism will focus on becoming a Lean Service Provider, Infrastructure Factory, or Application Integrator.*

Before you start reading, please bear with me. Because I tricked you: I'm not going to talk about "mature and emerging telecoms". Well, I do, but maybe not the way you would've expected me to. Let me explain.

What is an Emerging Telecom?

When asked about emerging telecoms, most people probably think about African countries, maybe an Indian Competitive Local Exchange Carrier (CLEC), or rural providers in the BRIC countries – Brazil, Russia, India, and China. However, when evaluating the acceptance of a new way to consume IT services – the Cloud way – it is helpful to take a close look at ICT adoption rates. The Intel researcher and anthropologist Dawn Nafus explains:

> *Managers in technology firms often say to us that of course mobile phones have taken off in emerging markets, as building the network is so much less expensive. Yet the reduced cost and lack of path dependence is not a sufficient explanation for why and how this has happened. For example, it does not explain right away why it is that people in the region often purchase mobile phones before they do refrigerators; or why they purchase handsets which analysts would predict to be 'unaffordable'; or why the role that user-driven innovation and how larger actors responded to those innovations had in making those technologies relevant and meaningful in those markets. In fact, many 'unaffordable' technologies, such as solar panels, can be found in abundance within low-income parts of the world: what is an expensive, nice-to-be green accessory for the rich can be a mission-critical infrastructure elsewhere.*

Dawn Nafus, Phillip Howard, and Ken Anderson subsequently created the Technology Distribution Index (TDI)(Nafus, Howard, & Anderson, Getting Beyond the Usual Suspects: Policy and Culture in ICT Adoption, 2009), a calibrated measure that allows us to relate a country's share of the global stock of a particular technology to its share of global GDP dollars. The TDI is calculated across four technology categories – mobile phones, computers, Internet users, and Internet bandwidth –to say whether the country has more, less or a middling amount of technologies for its level of economic capacity.

While the datasets required to calculate the TDI are not always reliable, it opens up a new way of thinking which countries might be quicker in adopting Cloud services than others, which telecoms might be operating in emerging environments and which ones operate in mature environments. Dawn Nafus then went one step further to derive the Technology Metabolism Index (TMI) (Nafus, An Anthropologist's Eye for the Tech Guy: Emerging Market Opportunities in a Post- BRIC World, 2010), an ethnographic and statistical model that tells us which countries have out-performed the average rate of global technology diffusion, given each country's economic productivity. The outcome is astounding. The top 15 countries with a high TMI score are (in that order!): Estonia, South Korea, Malaysia, Dominica, Moldova, Kenya, Morocco, Turkey, Benin, Paraguay, Bulgaria, Slovak Republic, Lebanon, New Zealand, and Israel. Her key finding is that high scoring countries – both rich and poor – share cultural factors that make them rapid tech adopters:

- The countries all have a strong sense of "we", a strong nation-state, with an external economic competitor to measure against. For Estonia the competitor is Russia, for South Korea it is Japan.

- The countries all have a "normative ethos": New products are perceived as 'normal' before they become ubiquitous. Highly innovative products are often re-coded as something appropriate for ordinary people, not only for the wealthy or high-income class.

- The countries all have a high social and physical network density; such networks are non-linearly harder to build in larger places.

- The countries are all in an active and agile state – not necessarily democratic. Technology is in fact often seen as part of a clean slate approach after radical social change.

The clear lesson learned is: rapid technology adopters are shaped by cultural ecosystems and public policy environments, and not by the individual buying power and decisions. As a result, the distinction between emerging and mature markets does not seem helpful. In fact, the traditional radars for where to look for partners and markets will lead to missed opportunities. Telecoms should be aware of such differences in technology metabolism, and we will suggest different strategies for such countries and their operators.

When discussing these findings with telecoms in the US and Europe, some jumped to the premature conclusion that either the technology uptake rate in these regions was so high because they had to "catch up" to meet the "standard" of "countries with a saturated technology ecosystem", or that people in these regions were buying "last-generations technologies" that first-world innovators are dropping from their portfolio and now exporting as a second life cycle. There are two arguments against that premature conclusion. First of all, Dawn already shows that the technology uptake rate is independent from the "degree of technology newness" (for the lack of a better term). And South Korea is probably snickering right now about broadband uptake rates, network innovations, and digital services in the US and Europe. I don't think they have to "catch up" to anything.

The second argument is a lesson-learned for telecoms. Which innovation is more beneficial to your end customers: A cell phone that can run for two weeks without charging and allows secure financial transactions and notifications over extremely low data rates; or a cell phone that requires a charger every 12 hours, with fragile and hard-to-read-in-burning-sun touch screen, requiring a complex infrastructure of app stores and billing systems, with lots of partner and vendor relationship management and content ownership discussions? I bet in some countries you would opt for the first, in some for the second, depending on your customer needs. Aneesh Chopra, Federal CTO of the United States, hits it on the head:

> GE has introduced medical diagnostic equipment at 85% of the price point, which is profitable in rural China. We can re-import that technology and innovation into the US domestic market to expand the marketplace – redeployed to the US with more modular design, more sharable intellectual property. I see tremendous opportunities looking at emerging economies as target markets for US firms wishing to exploit principles of reverse innovation: Innovate in these countries where price constraints are such that traditional services are too costly. Gather that feedback and learning, and re-import that feedback as GE describes. I think that same play will be played out in the US technology sector. And if we don't play it, others will.

I was discussing with Dawn if there could be a common denominator for telecoms, a common characteristic in countries with higher technology metabolism as well as for countries with lower technology metabolism. We found a few interesting hints on these characteristics, despite massive regional differences in culture, networks, regulation, legislation, and GDP:

- One could interpret the TMI as an index for spending on digital services and technologies compared to the GDP. As a result, in countries with a higher technology metabolism the Average Revenue Per User (ARPU) percentage of income spent on digital services will be higher than in countries with a lower technology metabolism.

- Telecoms in both countries with higher and lower TMI are battling costs of operations and capital expenditures. Just because a telecom in a country with higher TMI has a relatively higher ARPU percentage of income on digital services does not mean that it is more revenue structure focused than telecoms in countries with lower TMI. Both telecoms may operate in environments of fierce competition where price matters, and hence production cost matters.

- A higher TMI – or higher ARPU percentage of income spent on digital services and technologies, for that matter – does not say anything about the level of competition for digital services ARPU, nor does it say something about the level of innovation at telecoms. Kenya's mobile carrier competition is as fierce as in the US. The level of telecom innovation – productization of inventions – in South Korea is as high as in Germany.

- Telecoms in countries with a higher TMI have to tap and defend new markets aggressively. The challenge is not so much tapping new ARPU as keeping it with agile new service offerings, often resulting in stove-pipe-like launches of otherwise mostly independent digital services. Telecoms in countries with a lower TMI are focusing on solutions, carefully evaluation how to stretch into higher ARPU – and whether a new service or feature is actually tapping new revenues or just cannibalizing revenues from existing services.

- Telecoms in countries with higher TMI operate in an environment where a somewhat open commodity exists as an enabler to reach – theoretically – all people. In South Korea it's the relatively cheap and abundant high-speed Internet access, creating myriads of new digital services and products. In Kenya the enabler commodity is the mobile networks in a highly competitive market.

One approach to sketch out a mid- to long-term strategy for telecoms is to no longer classify them as "emerging" or "mature". These terms will be pointless in ten to fifteen years, when a second wave of innovation, an economic downturn, a market shakeout, consolidation, or fruitful co-opetition could have changed their strategic positioning anyways. Instead, telecoms could derive their strategies according a region's technology metabolism they operate in.

Strategic Impact on Telecoms in Countries with Higher Technology Metabolism

Coming back to the six future cloud business models we discussed (see picture below), telecoms in countries with a higher technology metabolism will open up great opportunities for differentiation, given the speed of technology adoption.

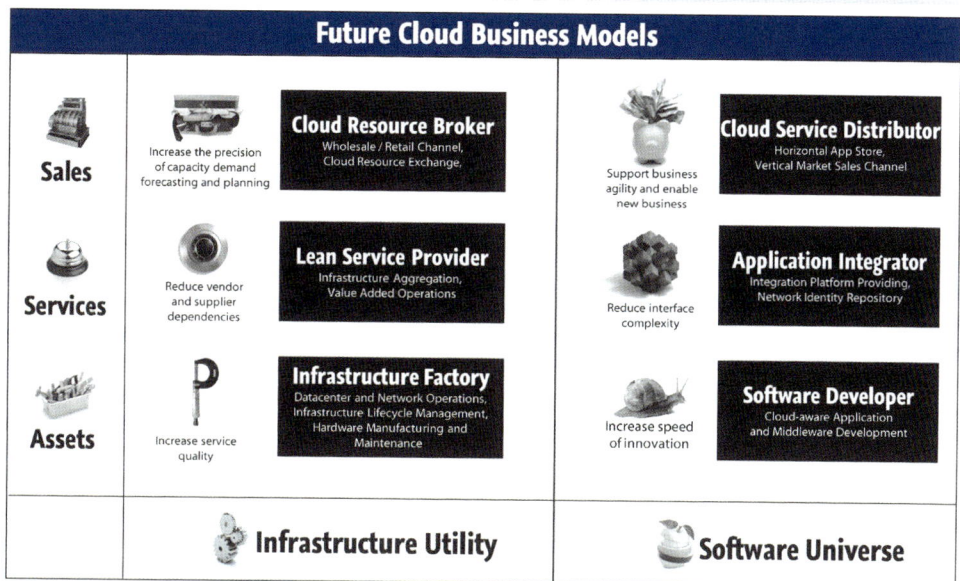

Future Cloud Business Models		
Sales	Increase the precision of capacity demand forecasting and planning	**Cloud Resource Broker** Wholesale / Retail Channel, Cloud Resource Exchange,
	Support business agility and enable new business	**Cloud Service Distributor** Horizontal App Store, Vertical Market Sales Channel
Services	Reduce vendor and supplier dependencies	**Lean Service Provider** Infrastructure Aggregation, Value Added Operations
	Reduce interface complexity	**Application Integrator** Integration Platform Providing, Network Identity Repository
Assets	Increase service quality	**Infrastructure Factory** Datacenter and Network Operations, Infrastructure Lifecycle Management, Hardware Manufacturing and Maintenance
	Increase speed of innovation	**Software Developer** Cloud-aware Application and Middleware Development
	Infrastructure Utility	**Software Universe**

Differentiation needs more than commodity infrastructure. High metabolism means fast adoption, means great opportunities to differentiate on the application and sales frontend. The high technology adoption will also create a fast-paced information and communication technology ecosystem of smaller players in verticals. Digital services will come and go at a pace beyond telecom's ability to create a stable long-term infrastructure factory. The Cloud services in demand will change too quickly with the technologies and services of consumers and businesses.

But telecoms can leverage their vertical market access, know-how, and scale of operations in a market that demands increased precision of capacity demand forecasting and planning. Becoming a Cloud Resource Broker is a viable telecom business model in the infrastructure utility segment. Just as entrepreneurs and startups need a trusted single point of contact for their changing computing demands, they need the reach of a telecom while maintaining business agility. Becoming a Cloud Service Distributor with horizontal App Stores and vertical market sales channels is another viable business model for telecoms in countries with high technology metabolism. Last but not least, while platform buy-in and gravity might rapidly change the need for higher speed of innovation remains. The development of cloud-aware middleware and applications will be essential for new players to increase their speed of innovation.

Of course making strategic recommendations for any kind of telecom without knowing their focus, competencies, and specific environment is pointless. But we can give you a rough idea on the strategic impact in three important categories organization, integration and partnering, and services.

Organization

So what people and which processes are needed to leverage the full potential of Cloud in the coming years? The task seems challenging when looking at some of the constantly changing internal and external factors: user needs, internal stake holder requirements, technology and vendor developments, regulatory boundaries etc. Handling all these factors is and will be a major task for any IT organization within a telecom. In addition to that, internal IT departments will change dramatically through introduction of cloud technologies.

Yousef Khalidi from Microsoft elaborates on these effects: *"Look at what cloud computing gives you. It frees you as a customer from having to deal with the low level stuff, such as provisioning machines, bringing the machines up to speed, booting the OS patching, running the network and so forth. These functions, by and large, are performed by people today and those people effectively have to move up the stack and spend more of their time running the business and running the applications above the infrastructure instead of dealing with the infrastructure. Especially the job roles that exclusively deal with maintaining infrastructure will be substituted by automation. And this will cause some resistance."*

Any established IT organization that is seriously trying to position itself as a leader in adopting Cloud business models and technologies will therefore feel significant pressure. The more that organization is trying to manage IT "the old way" the higher the pressure will become. And the pressure will come from both sides, from within the organization, where people feel the pressure to either change or get redundant, and from the outside, where a market with constant change and a number emerging players makes it difficult to navigate the supplier-ecosystem effectively. Key questions here are:

- Can I do this with my existing organization, or do I have to build a separate organization to manage this services portfolio?
- What level/amount of services orchestration do I keep in-house?
- How do I differentiate myself from potential competitors who are trying to launch similar services?

Internal Cloud initiatives need executive sponsorship – but rather from the CIO than from most technology oriented CTO. Jamie Allen from Net Service Ventures explains: *"Most people would say that cloud computing is not a technology, it's a business model. But underneath it, there are technologies that enable the business model. And that's what I think the internal IT, the CIO needs to focus on. The CIO gets rewarded for providing IT capability at the lowest cost possible, and this is a way of doing it."*

Integration and Partnering

All three ascendant business models for telecoms in countries with higher technology metabolism – Cloud Resource Broker, Cloud Service Distributor, and Software Developer – require a constant re-iteration of negotiating and maintaining partner contracts and monitoring systems integration according to defined Service Level Agreements (SLAs). While Application Integration is not a specifically dominant business model to sell to external clients, the internal application and interface integration efforts will be immense simply because of the high level of innovation and shifting powers in a relatively cash-rich environment (rich compared to a countries productivity).

As a result, one of the most critical success factors for a competitive cloud portfolio is finding the right partners to build and run these cloud applications. Yousef Khalidi from Microsoft sees partnering as one of the key components of success: *"carriers have to understand with which software companies they have to partner to get their solutions out. And they have to partner with companies that have partnerships as part of their business DNA to get the job done. So they need to identify the software companies that are leading in the cloud space and partner with them".*

He also sees partnering as one of the key weapons against commoditization and the threat of being degraded to running only the lower, network centric stacks of the cloud ecosystem: *"Unless telecoms play in the market, by partnering, by becoming the service provider, it will become more of a commodity play for them and they will just end up providing the basic pipes. If you look at the stack for cloud computing, at the very bottom you will find datacenters and the usual infrastructure. Then come the pipes and above the pipes there are elements like computation, storage, content distribution and you have services on this layer as well, reaching all the way up into verticals and given geographic locations and so forth. Carriers need to make sure to not stay on the commodity play level - they need to move up the stack a bit. How high they should move up the stack should be cautiously considered, especially with regards to investing into software, which is why I mentioned that it is important to partner with software companies that have the expertise and software stack to make cloud computing a possibility"*

But partnering isn't easy because of different sales and innovation cycles. Especially service and technology providers will tend to partner with already established telecom equipment providers rather than with telecoms directly to leverage a greater scale and reach. There will be strong competition from otherwise friendly allies to protect and extend value chain powers. *"Telecom related or oriented companies always have the same problems of sales cycle. This is particularly problematic in the U.S. You're either selling to the large telecoms, Verizon, AT&T, Comcast, Time Warner, et cetera, in which the sales cycle can be 18 months or longer. And you have to first fund the development of the company. And then you're basically funding this potentially never ending process of sales, which consumes a lot of capital"* Says Brian Wilcove at Sofinnova. *"My advice to companies that are in that space - and I have several - is usually to partner, and partner early. To defray sales costs and to try to shorten that sales cycle. So partnering with and Ericcson or and Alcatel or a Cisco, whoever it may be. Because by and large the telecom equipment manufacturers are turning into more and more service oriented companies. You have HP and Oracle on a traditional server and compute side trying to get into that business. You have Ericcson and the same basically wanting to do more and more operations. Selling into those channels is becoming more and more important."*

Partnerships will shorten integration cycles, but also give customers a "reason to believe" that the agile and high-quality technology adoption will actually take place at the telecom and is no marketing ploy. Transparently showing investments and partnerships publishing is one way to do that. *"We made major investments in the last year. We made acquisitions in the areas of a data mining. We made enormous improvement to our organic analytics capabilities. We have very strong partner relationships with companies like Informatica and Pervasive and Cast Iron and Tibco whose whole expertise is in integrating data, repurposing data, and finding ways to get more business value out of existing data."* Says Peter Coffee from Salesforce.

A well designed partnering strategy requires a clear understanding of the roles of each partner, a dedicated partnering organization, as well as effective governance and operational processes. What organizational function should this partnership organization report into? Experience has shown that only a direct reporting line into the top management/CEO makes a strategic partnership organization effective, as well as local partner management.

Services

What are the characteristics of services in high technology metabolism markets?

- High-speed launch of turnkey solutions: Higher technology metabolism regions show very competitive service landscapes for consumer and business wallet share. Speed of implementation is one of the most important success criteria to create early buy-in and stickiness in an otherwise fast-churning market.
- Management of large amounts of low complexity solutions: high metabolism and adoption at the consumer and at businesses create opportunity for a large amount of applications and services

- Low level of integration but high level of scale and application performance: a fast launch of individual services will lead to low levels of integration, however, as the uptake of services can be quite significantly, it is important that services can scale quickly
- Quick phase in/phase out of products. A high trial and error rate is necessary to keep the amount of innovation and rate of successful services at a good level

Services in regions with higher technology metabolism will be the test bed for other markets around the world. As innovation is often driven through proximity to a large amount of early adopters, these markets will act as early innovators of services.

Strategic Impact on Telecoms in Countries with Lower Technology Metabolism

Countries with lower technology metabolism have a lesser spending on technology and digital services compared to the country's productivity, its gross domestic product (GDP). Differentiation is difficult. These are often highly fragmented and saturated markets. As a result, service providers in these regions will often compete on price rather than features. It requires extremely cost-effective production to squeeze the margin; and highly integrated solutions that prevent customer churn, allow micro-step up-selling, and organically creep up the value chain and increase ARPU share. Both capabilities are neither cheap nor easy to build and also require careful planning of service offering portfolios to avoid premature ARPU substitution and service cannibalization.

At many telecoms in regions with lower technology metabolism we see the desperate – and difficult! – search for new service areas and "untapped markets" that justify the search effort. To make matters worse, these telecoms are often already active in many business and consumer industries and services, and the net present value and effort to find incremental revenue is higher than the revenue itself (or an investment into money markets). I was in the fortunate position to work with some of the smarter telecoms like Etisalat, VNPT, or Deutsche Telekom Group as well as service providers like Microsoft, Yahoo, or Google to end this vicious cycle in saturated markets with low technology metabolism. They were focusing on large-scale and open enablers that will ultimately fuel the technology metabolism, also in secondary markets of suppliers and vendors.[9]

These three market drivers – cost-effective production, highly integrated solutions, and large-scale and open enablers – are great opportunities for telecoms in three Cloud business model areas:

---

[9] Strategically this is a tricky problem: How do you animate consumers and businesses to spend a larger expense share on technology and digital services, while at the same time you maintain a positive impact on the quality of life? As a telecom you don't just want to sell more, your promise to shareholders is mostly a solid, steady-as-she-goes value stock combined with a trustworthy stable brand image. You don't want to undermine a long-term business strategy with luring customers into services they don't need.

- Lean Service Provider: Provide infrastructure aggregation and value added options to reduce vendor and supplier dependencies;
- Infrastructure Factory: Datacenter and network operations and infrastructure lifecycle management to increase service quality;
- Application Integrator: Provide integration platform and network identity repository to reduce interface complexity

All three are enablers for third parties as well as for internal efforts, and all three leverage a telecom's ability to accomplish a large, debt-funded infrastructure refresh. Let's have a look at the same topics we discussed for high technology metabolism countries.

Organization

The organizational requirements for a telecom in a lower technology metabolism region will look quite different from the ones we discussed for higher technology metabolism regions. As carriers in these markets are highly focused on enabling services and integration, a large amount of infrastructure vendors have to be managed. This is closer to what carriers are doing already in their day to day business. The core component that is missing here is software competence. Most of the carrier IT organizations today remain highly networking technologies centric. However, to be able to compete in the cloud infrastructure game, deep expertise in selecting, developing, managing and maintaining this infrastructure is necessary. Today, the lack of software skills is often mitigated by a large amount of outsourcing. This strong dependency on vendors leads to a lack of differentiation potential, and also to strong dependencies that reduce the level of flexibility for future decisions.

Integration and Partner Ecosystem

As discussed above, in order to cover the lack of software skills and to reduce the risk of having to manage a higher degree of complexity, some carriers will take advantage of implementing and running turn-key solutions from large single vendors. The flip-side: turn-key solutions can, under certain circumstance, cause a vendor lock-in: This is where Richard McAniff from VMware sees the advantage of virtualization: *"[Telecoms] want to be able to have choice as they pick the appropriate cloud. They want to be able to say 'I don't want to get locked in'. Part of what we do to is to make sure that we are able to provide bridges to these different environments for the enterprise itself, so that if you want to run different kinds of clouds we need to make sure that we are providing that ability. In terms of the solutions providers, I think it is important to provide them with some infrastructure that they use, but then you have to allow them to extend that infrastructure as they see necessary. And ultimately they can choose to be a development shop, they can leverage software micro-cells and at the end of the day, if you are using our software, they might want their own data bases."*

Effectively managing this partner ecosystem will be a very important task which requires a thorough understand of the partners capabilities, future roadmap as well as financial stability. *"What we'd like is the likes of NTT and Deutsche Telekom and maybe Singtel to partner and give me a single face, a single interface, single pricing, or at least single price list, where I can now go to a customer like Deutsche Bank and say, 'Here you go. I've got 256 points of presence around the world in these countries, and essentially we can go wherever you do business.' That would be nice."* Says Ishmael Ghalimi from Intalio.

We don't believe that there will only be a handful global players after a massive consolidation. *"I do not think that there will only be a handful of providers across the globe for a couple of reasons: One is the fact, that we do have national boundaries. We do have government regulations and I do not believe that governments will be happy with having a few mega-companies controlling computing for the whole globe."* Says Yousef Khalidi from Microsoft. *"I do believe there will be consolidation in the space, perhaps on the low-end hosting, which will be consumed by higher-end hosting. But the number of competitors will not be low in my opinion. It will perhaps be several hundred in the telecoms arena and some longer tail for certain locations, verticals and countries, coupled with the fact that I also do not believe that we will walk away from on-premise computation. You will always have an on premise presence that will have to be managed through more traditional means, coupled and integrated with a number of Cloud providers. And that number, as I mentioned, I believe to be in the hundreds and not just a handful."*

Services

What are the characteristics of telecom services in low technology metabolism markets?

- Strong focus of carriers on enabling services: low technology metabolism regions will most likely lead to a strong focus on enabling technologies, as described above, leaving the field for external innovators to build customer facing services on top of this infrastructure

- Highly integrated Services: customers will be able use a number of features and highly integrated products, making it easier for market participants to absorb these service but harder to switch telecoms and thus prevent churn and capture customer mindshare.

- Large, complex services: the focus on only a couple of "blockbusters" will give room for the development of large, complex services. Differentiation is created through high barriers of entry.

A final word on the different cloud business models we discussed. Which one to pick will be a challenging decision in each case. Given the size and competitiveness of the market it will be very difficult to compete in more than three areas. Focus on core competencies and target customers should be the starting points on this journey.

Interviews

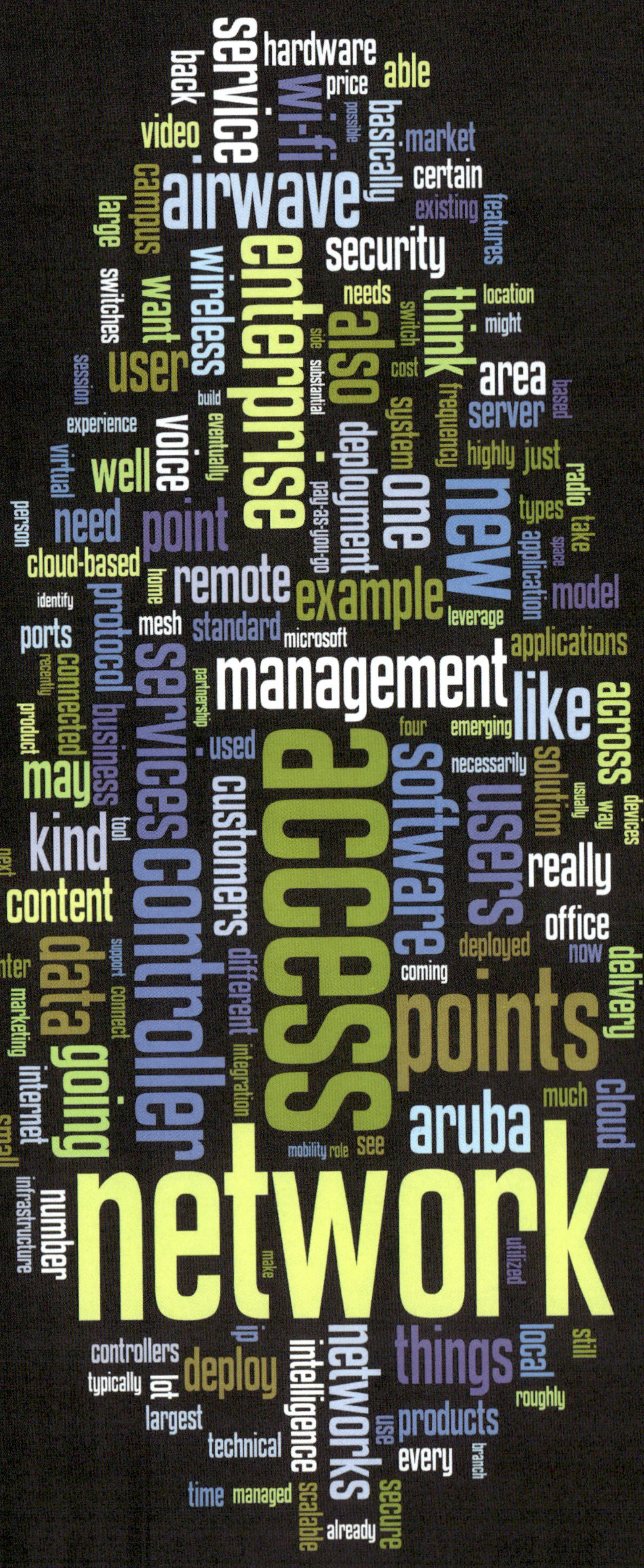
network
access
points
management
controller
services
enterprise
airwave
service
wi-fi
hardware
able
back
price
video
basically
possible
market
certain
campus
large
existing
features
security
wireless
want
switches
needs
think
location
might
also
switch
side
substantial
user
cost
area
session
build
frequency
server
based
experience
system
eventually
deployment
voice
well
virtual
highly
just
one
new
radio
point
person
need
pay-as-you-go
types
take
cloud-based
remote
example
application
space
home
identify
protocol
mesh
model
standard
leverage
ports
microsoft
applications
recently
connected
partnership
like
across
product
business
four
emerging
used
may
software
necessarily
solution
users
usually
way
devices
kind
customers
tool
next
really
content
start
data
office
deployed
now
support
different
delivery
center
coming
connect
called
marketing
going
integration
terms
internet
much
cloud
aruba
small
mobility
role
see
number
infrastructure
make
utilized
ip
still
controllers
things
local
typically
lot
deploy
networks
intelligence
roughly
largest
use
products
technical
every
branch
time
managed
scalable
already
secure

Aruba Networks
December 1ˢᵗ, 2009

Within the past few years Wi-Fi networks became a pervasive commodity within the enterprise. But last year marked a turning point for three reasons: First, the much faster 802.11n standard got ratified in September 2009, and with it more data hungry applications and services will burden the network. Second, the economic downturn made many companies realize that the previous over-provisioning of Ethernet ports created prohibitive costs within the wiring closets for maintenance, software upgrades, and refresh for those ports. And third, as applications, services, and maintenance move into the cloud, so does the network – and with it the management of its capabilities.

In mid January 2010 Detecon's Thorsten Claus met with Aruba Networks' Head of Technical Marketing, Paul Curto on their spacious Sunnyvale campus. The open office space is bustling with engineers huddling over blueprints and pointing at exposed electronics of prototypes with cables hanging out. But more so, software developers are scanning through code, and a whiteboard is densely scribbled with schema and database information and event flows.

THORSTEN: Hi Paul, great to have you. Why don't you start with telling our readers a bit about what you do?

PAUL: I'm Head of Technical Marketing at Aruba Networks. I'm basically responsible for competitive analysis; trainings for the technical field, which is our technical sales and engineers; application notes and technical briefs on unique features about our products and how to use them. While we're part of Marketing we like to cut through the high-level marketing messages down to the very deep technical aspects of operations, applications, and benefits.

THORSTEN: By now most Internet users have a Wi-Fi router at home, or know how to use a Wi-Fi hotspot in a café – you might have to enter a security key, you might get a login screen once you fire up your browser, etc. What's so different between a consumer solution and Aruba Networks' specialized enterprise mobility solutions, enabling secure access to data, voice, and video applications across wireless and wireline enterprise networks?

PAUL: Let's make sure we are talking about apples and apples: When you typically think about a user experience of Wi-Fi, you are thinking about a consumer experience at home, or a small office experience. However, typically in most of the enterprise environments that we work with, an IT department may be responsible for setting up your system, setting up the applications, setting up the identification credentials, up to a point when they hand you your enterprise-class laptop.

What our solution is targeting is the enterprise Wi-Fi, corporate IT department, the manager of IT – the person who needs to make a decision about investing in Wi-Fi infrastructure. We go as high as the CIO-level. We have to also target the CSO-level and really kind of calm or relay to their concerns about security. There is sometimes the notion that Wi-Fi is inherently unreliable, unsecure – even though one could connect a [Local Area Network] cable in any conference room and very likely get access to a whole flood of unencrypted IP packages on that segment. With Aruba what we offer is a highly scalable, manageable, reliable, and secure enterprise class product. If you want to deploy a pervasive network across the campus this is a much different IT challenge than deploying a single access point in a home environment, where what you really only need is basic access and connecting to a DSL router …

THORSTEN: So how large are your installations usually?

PAUL: They vary in size, the largest installations being in excess of 10,000 access points. Microsoft, for example, falls in this category for large enterprises. We also have educational customers, such as Ohio State University, which is the largest campus in North America. There are also very large deployments in New South Wales in Australia, for example. We had a press release on that in partnership with IBM, where I believe they have 250,000 students, each with their own laptop. And then you have classrooms with a high density of users, 40 or 50 users accessing the network at the same time. So the Wi-Fi needs to be highly scalable in very localized wireless environments like a lecture hall, but also very scalable in terms of the number of sites. We also have clients in large hospitals and retail. One of our showcase customers is Safeway.

Just to give you a feeling for the deployment schedule: there are 1,800 stores that were deployed in roughly three to four months in partnership with Orange Business Services. Because we have a very scalable, easily deployable solution we were able to do that kind of rapid deployment, bringing everything up and running and getting it ready for the store openings …

THORSTEN: There must be lots of issues with management, security, deployment, processes, with authorization of devices, with shutting down compromised or rogue devices, all of these things. That sounds fairly complex.

PAUL: That's accurate. And what you have to think about, too, is a lot of our customers have existing backend infrastructure that they want to integrate with. What we have is an Aruba Networks mobility controller, a wireless [Local Area Network] controller. Our largest controller has a chassis with four modular blades that go into the chassis. Each blade can support up to 512 access points. So you have a total of roughly 2,000 campus connected access points per controller. There are a number of things the controller can do. Some of the primary ones are a stateful firewall that is integrated in the controller. We have a user identity-based management system: When we identify users, the first thing we identify is who is that user, what role does that user have, what services, what network access, what resources is that user allowed to access. The backend systems might be, for example, Microsoft Active Directory with [Microsoft Internet Information Service] as a frontend, so we can do Radius integration, we could do [Lightweight Directory Access Protocol] integration. The controllers serve as kind of a proxy for these types of messages. The Radius server may return an attribute saying *"this user is in this role"*: This is a student, a contractor, a guest, an employee, or a marketing person. Depending on the role attributes that are returned, the controller will drop that user in a particular Virtual Local Area Network (VLAN) or assign particular access lists and stateful firewall rules specific to that user. It is able to identify that user to the system and that profile follows the users throughout the system, defining where they can act and how they can act.

THORSTEN: I remember four or five years ago everyone was talking about distributing more intelligence to the edge. But what you are describing seems like you are starting to consolidate a lot of intelligence. Is that because price per access point is critical at a deployment of 2,000 access points? Or were there other drivers involved?

PAUL: Price is a driver, although access points by themselves … it's a highly competitive industry: the prices of access points have been coming down for both the legacy and new products. When 802.11n, the new standard, was announced, there was a first draft of the standard and a lot of the enterprise vendors started releasing products right away at much higher price point. We've already seen roughly half of the price points after only a few years having that draft standard. Now it's a fully ratified standard. And the price point of the actual hardware tends to be dropping. Year after year, there's next generation silicon. The 802.11n standard has optional parameters that you can implement, and as the new silicon vendors started to employ those new optional features, that helped do differentiate a little bit. But the real differentiation is in the management software.

That comes back to our controller models. We are making the life of an IT person easier by allowing small teams of people who have some knowledge of Wi-Fi without the need of a [Radio Frequency] Ph. D., to be able to operate large scale enterprise class networks and just deploy the policies that they want on the infrastructure they have the way that they are used to, like for example, their Radius server, their [Lightweight Directory Access Protocol] server. We enable them to leverage a product like Aruba's controllers to make the deployment of access points very, very simple. The access points tend to be lightweight access points in a sense. There's not a lot of intelligence that the access points have a priori. And then the access points essentially build tunnels back to the controller.

Controllers are typically deployed in one of few different modes. There is a campus connected mode, where the access point is connected across a [Local Area Network], so the controller in that mode usually uses a GRE-based tunnel[10].

We have a remote access point which is typically deployed across a [Wide Area Network]. For example, if you have hundreds and hundreds of stores in a retail environment and you don't want to deploy controller in every store, you could deploy a remote access point in every store. And then a secure tunnel is built back to the controller using IPSec for security[11].

---

[10] Generic Routing Encapsulation (GRE) is a tunneling protocol developed by Cisco to encapsulate a wide variety of network layer protocol packet types inside IP tunnels.

[11] Internet Protocol Security (IPSec) is a protocol suite for securing IP communications by authenticating and encrypting each IP packet of a data stream. IPsec also includes protocols for establishing mutual authentication between agents at the beginning of the session and negotiation of cryptographic keys to be used during the session.

We also have mesh-based access points that can operate as a "mesh-portal" or "mesh-point"[12] and be able to build networks out where you may have difficult access to cabling and so forth.

From the cost standpoint the hardware costs are coming down, so margins can be maintained, while the controllers are becoming more and more scalable. It really comes down to the management and software of the network to enable offloading the IT person, so they don't have to spend a lot of time troubleshooting connectivity problems. They just deploy the access points where they need the coverage and then they don't worry about it. And then, of course, they have the management tools, dashboards, troubleshooting, monitoring, and visual tools in order to troubleshoot the network.

THORSTEN: So the network infrastructure business is accelerating towards a more software driven model. Is that your experience as well? Is more and more intelligence going into Cloud? Are we now seeing the next generation Network-as-a-Service or something like that emerging?

PAUL: I think that's definitely an emerging trend. There's a company called Meraki who used to be the MIT Roofnet Project. They have kind of a mesh-based architecture, but it is more about the business model that's innovative. They have kind of a pay-as-you-go type of a service that is managed in the Cloud. More recently, you can also take our AirWave product, which is our multi-vendor network management tool. Aruba purchased AirWave back in early 2008. They were one of the very few tools out there that did multi-vendor wireless network management.

Just for scalability discussion: Cisco's largest wireless [Local Area Network] deployment is the New York Department of Education. And Cisco's largest wireless [Local Area Network] network is managed by AirWave. That provides an indication of how well AirWave has done in terms of integration and terms of the features and ease-of-use that customers require to manage other networks. But that level of integration, of course, is now there as well within Aruba. There's roughly 16 other enterprise wireless [Local Area Network] vendors that can be managed by AirWave. AirWave also allows customers to basically defer upgrades: if they have an existing Proxim or Cisco network and they want to choose Aruba, but they don't necessarily have the budget to do a complete rip-and-replace, they can install AirWave, start monitoring their network, or do things like gaining PCI-compliance [Payment Card Industry (PCI) Data Security Standard] to certain types of reports. They can do things like locate and track rogue [Access Points] on their network. They can also protect their existing investment and migrate existing [Access Points] when it makes sense for them to do so, deploying Aruba only in certain locations, for example, but eventually rip-and-replace everything.

---

[12] Mesh portals use wired Ethernet for the northbound interface, while mesh points use wireless for their northbound interfaces. Multiple mesh points form nodes in the mesh network and mesh points backhaul their traffic back into the secure network.

AirWave has come out with a new service called AirWave-on-Demand. We had beta trial last year running where basically the customers would simply sign up at airwave.com and they would – with the signed beta agreement –copy and paste their [Virtual Private Network] setup that they were given into their controller and login to their personalized AirWave site, so everything is basically managed in the Cloud. And AirWave has their management platform. They have this software module called "Virtual RF" which is a [Radio Frequency] virtualization and location services. And they have a tool called "RAPIDS", which is a rogue [Access Point] detection and [Intrusion Detection System] event management tool. What this allows customers to do is deploy AirWave without the need to buy any servers, install any software, or worry about power, cooling, maintenance; they don't have to install any software patches or upgrades, or worry about backups. It is a pay-as-you-go-service. Start using it and see if it meets your needs. And also, if your desire is eventually to have an AirWave server: there is an entry barrier to some customers, especially smaller customers who don't necessarily want to buy a server and host it themselves.

They are starting small with this pay-as-you-go cloud-based service and then eventually convert it to a service that they host and manage themselves, without losing any historical data. So it's a pretty cool tool and the service has just launched recently. All you really need is a [Virtual Private Network] connection back to the AirWave hosted location from your network where you have, let's say, an Aruba controller that is [Simple Network Management Protocol] deployed.

THORSTEN: When I was visiting your location it became very apparent from the workspace that this new emerging market is very different from what network equipment manufacturers used to do: I saw extreme programming – buddy programming –, I saw an open workspace with long tables, no cubicles, lots of toys lying around, lots of IT and hardware, everyone was very enthusiastic, programmers huddling together – this was mayhem and rapid development in the best possible sense. It is very, very, software centric. How does Aruba Networks – coming from the hardware and controller side – deal with this change as an organization?

PAUL: I think Aruba started out as a company that was focused on both hardware and software from the beginning. As we move forward in the market, there is still kind of an equal focus on coming out with new hardware, with new platforms, with new products that meet certain market needs, as well as optimizing software and continuing to add new features. It's always really been kind of a hybrid approach, to be honest with you. For example our Virtual Branch Networking, which is our remote access line of products: It really was more or less an idea that I believe our CEO had, socialized with one of our key customers, and he basically said: *"Wouldn't it be a great idea to take our controller and leverage what you already do and use that for remote access solutions, remote network solutions?"* And within a six month timeframe of that discussion, there were five new products that were rolled out, that ranged across the remote teleworker with very small enterprise class 802.11b/g access points for less than a hundred dollars – a very cost effective solution – up to larger branch office controllers that can do things like network storage, services, print services, USB for 3G remote uplink for site survivability, and so on and so forth ... so products for a range of different sized offices, from a remote teleworker up to a small and medium brand office. There needs to be a focus on hardware to be able to accelerate new platforms to market, while at the same time there has to be a focus on software to be able to leverage the innovation we already have and then attack new markets in build, new ecosystems, as we continue to expand our product portfolio.

THORSTEN: It sounds like your software centricity enables you to innovate much faster and roll out new features much faster. I recently read about Aruba's high quality voice and jitter-free streaming video. Configuring that with 2,000 access points was before nearly impossible, business-wise, because you had to touch every single base station.

PAUL: Two things to the cost point: One is we have a program called "Network Rightsizing" where we basically look at our existing customer revenue prospect and say *"Mr. Customer, let us help you figure out what percentage of your edge switch ports are actually being utilized today."* And it could be anywhere from *"You deployed four ports in every office, in every cubicle, and only two are being effectively even connected to something, two are just sitting there, idle, nothing connected in"* ... Well, you could have some users connected into the switch, but they're only surfing the web and doing some e-mails.

What you find is, as you do an analysis of your network core utilization, that there are a substantial percentage of ports that are either not being utilized at all or highly underutilized. For the ones that aren't utilized at all, consider that you not only paid to run wires to every one of them (which is sunk cost) but those cables still consume switch ports on the other side, in a wiring closet. Those switches usually have a fair amount of unused port capacity. You could consolidate all of the ports that are actually lit and being utilized into fewer switches and then take the users who are light users – who are running mission critical applications but are not necessarily that bandwidth intensive – and start migrating those user groups to primary wireless access. You may still have wired desk phones for Voice over IP that you still need to connect to a switch. You may have some other devices that need to connect in, you may have workstations. You may have heavy CAD software development typed users who need to have wired connectivity for some applications. But at the end of the day, you can still convert a substantial percentage of your users and decommission a substantial number of ports, consolidate those switches, and then take the savings you are already spending on closet refresh for your switches as well as savings from spending on your maintenance as well as your maintenance fees that you pay on your switches – and deploy a pervasive Wi-Fi network across your campus for those primary users.

Another brief point to follow-up on video that you've mentioned: Voice and video and other real-time applications ... We have really been spending a lot of time on the innovation side focusing on [Radio Frequency] management, on [Radio Frequency] optimization, on [Radio Frequency] performance, on combining that with our application intelligence, which is only possible with a controller, stateful firewall, and the [Application-Level Gateway] providing deep packet inspection that are built-in. We have the ability to detect whether something is a SIP-based communication and then say *"classify that as a voice call"* and we have the ability to detect a [Microsoft Office Communicator] video session that has been activated and can be classified as a video session. Then once you have that application intelligence and application awareness, you can now take certain actions on it.

We have, for example, at Liberty University a 15 HD-television channel deployment, pervasively across the campus, where we had to put special software in that did the optimization for multicast and multicast-handling reliably over Wi-Fi – a difficult problem to solve – mixed together with [Internet Group Management Protocol] support for IGMP, IGMP Proxy, IGMP Snooping. At the same time that same network is being used by faculty and certain select staff for voice. They have voice handsets and they have Cisco's IP Office to run SCCP-based voice calls. Then, of course, they have the data network for the students and the faculty. So that same Wi-Fi network is being used for a pervasive voice, video, and data deployment across the campus which is really pushing Wi-Fi to its limits, if you think about it.

THORSTEN: Are we going to see more and more centralized, cloud-based services? Is the network going to become more and more software dependant – which services run on which network or on which channel?

PAUL: I think there are a couple of different trends at play. Certainly having centralized intelligence is going to be helpful for certain things. And Cloud-based services are going to be one major trend, I think, management is going to be another. We talked about AirWave and how AirWave is leveraging Cloud-based services for multi-vendor enterprise Clouds, management over the Cloud. We could see, for example, service providers getting into this space because if you're going after especially small and medium businesses and you don't have a large sales force, AirWave-in-the-Cloud is one way to set up the service, market it, and push it out towards these types of businesses, maybe work through a value-added reseller type channel model and develop your business that way.

When you look at some of the other things, some of the other challenges to tackle, one of the areas to watch is application acceleration. We think that's a key area that will eventually require a solution. When you think about it in terms of content delivery there are some challenges around content delivery networks not being that enterprise-grade or -secure. You will see a combination of enterprise-grade security with content delivery with end-to-end encryption. You could leverage Cloud-based services, but you would have to have kind of a hosted model in order to have support for that.

THORSTEN: That's a very interesting thought. Who do you think are going to be the players in that space? Is it going to be the telecoms, the traditional [Content Delivery Networks], or the emerging players like Adobe that step up their game and start providing more enterprise-geared services?

PAUL: Today CDNs like Akamai, Limelight, CDNetworks, Amazon, or BitGravity others are not enterprise-grade.

THORSTEN: They might argue with you about that statement because distribution of movies and TV content is a huge industry and nothing to be trifled with.

PAUL: But they are not enterprise-secure and they are not enterprise-focused. They only deal with web based traffic they don't necessarily deal with dynamic data. A solution that can partner with those [Content Delivery Networks] in some ways, but also secure the data, is one that is going to be more targeted and more useful for the enterprise user.

I do believe Aruba will be one of the companies that is visionary in that space and will be well positioned over the next, I would say, the next couple of quarters. And we will see kind of how that plays itself out in the market as things are announced. Other areas that could be a of interest are things like the content security itself and how you can actually ensure things like URL filtering services, how you can figure out web policies you want to enforce for your enterprise users, add remote offices. Being able to deploy that in a central location is one possible solution, a centralized service that can then provide these services.

THORSTEN: So what you are saying is those kinds of services are also in a Cloud as well?

PAUL: That's correct. When you really think about the way that this model can work: you have basically an enterprise data center, where you are going to have your Aruba controller for example. You are going to have potentially some source data. You can have your routing connection to the Internet. The Internet then would connect into a Cloud based service. And then you could have branch offices, users at home, and then have a highly secured connection between the enterprise data center – which has the mobility controller – back to your remote office, and on the other end from the enterprise data center to the [Content Delivery Network] Cloud – or to the whatever it is in terms of the Cloud-based service you are looking for. It is kind of a two-legged approach where you have enterprise grade security across the entire end-to-end system, but you also have peering partnership or relationship in the data center in order to achieve that model.

THORSTEN: Isn't billing going to become a problem because you need to do some revenue sharing or develop new business and revenue models?

PAUL: That is a very, very interesting question. It may be done initially in a case-by-case basis. There are companies out there who may want to do a pay-as-you-go kind of model. They may have even OEM-types of agreement or other licensing relationships. The peering relationship has to have some kind of a business agreement. It could be transaction based, it could be the number of users, it could be the number of servers … For example, if you deploy a server in the enterprise data center that connects into the [Content Delivery Network] Cloud, that server can be sized based on number of devices supported, number of users supported – those types of things.

THORSTEN: Nice. You end up having a very granular business modeling for your clients: a pay-as-you-go service for short-term revenues; licensing for longer-term revenues; and bundling for up-selling, churn reduction, and the reduction of customer acquisition costs.

PAUL: You got it.

THORSTEN: Thank you very much for all the insights and for sharing your thoughts!

PAUL: Sure. It was my pleasure.

cloud services computing service clouds datacenter user able content capabilities ability set also going place people put new deliver mobility simple users new see fully different network virtualized applications service solution now cisco together end device problem just solutions power video experience network telcos access public models reduction networking enable way compute think devices demand less industry rapidly next-generation on-demand support architecture elements combine lot web enterprises markets look transfer number offering great higher seeing sure features move virtually traditional hardware mobile telecom growth cisco's big opportunity model build available processes refresh unified request critical space customer apis thing operate sps right get need like ensure virtualization external highlighted one execute change make structure looking connect whole well taking internal event heavy required rewriting around level opportunities things side delivery role thank key repurpose datacenters small times certainly time providers premium operators virtualize take rate really 20-80 come example enabling

Cisco Systems
February 2nd, 2010

Cisco Systems has a long history of working with telecom telecoms world-wide. As these telecoms move toward Cloud services and business models, so does Cisco. Cisco knows about the many competing goals at telecoms, from fixed and mobile traffic routing to high-bandwidth content delivery, and is taking a pragmatic approach to provide them with an end-to-end cloud computing solution: delivering Software-as-a-Service like WebEx, Platform-as-a-Service like Cisco's Unified Service Delivery, and foundational cloud infrastructure such as Cisco's Unified Fabric and Unified Computing System.

In February 2010 Detecon's Thorsten Claus spoke with Simon Aspinal, Senior Director, Service Provider Marketing. They discussed Cisco's vision of Cloud applications and services, and why device awareness and mobility will become a central part of a telecom's Cloud operations and offerings.

THORSTEN: Hello Simon, thank you for joining us. Why don't we start out with you explaining a little bit about your position at Cisco and what you're doing?

SIMON: Certainly. I'm the Senior Director for Marketing of Cisco's Mobility, Datacenter and Cloud Solutions to the Telecom and Media Industry.

THORSTEN: What strikes me is a title that includes mobility and cloud computing together. Why is that?

SIMON: These are both very interesting markets, evolving very rapidly and we are also beginning to see increasing cross-over between the two sectors. At Cisco we came to the conclusion that cloud is a critical element of the mobility sector and of course mobility is one of the key elements that enable the cloud computing sector... we see the synergies between both markets.

THORSTEN: So is the concept of cloud rather a new way to do mobile computing?

SIMON: Cloud computing is effectively rewriting the rules for a lot of user experiences, services, applications, and how they are being consumed. Mobility is rewriting the rules in terms of what you can access and where you can access it. We've seen an enormous plethora of mobile devices appear – smartphones, IP-enabled tablets, PCs, notebooks and now we're even seeing mobility linking machines and devices together. This change is being accelerated by the adoption of new services: cloud services and multimedia services being great examples.

THORSTEN: You describe a big topic for our clients as well. How is Cisco approaching these areas?

SIMON: The approach that we're taking as a technology vendor is that we have assembled a set of end-to-end solutions to address these opportunities. In the datacenter and the Cloud computing space traditionally the telecoms industry used datacenters for IT purposes, to execute applications and services internally and which would be consumed by the employees of these companies. 'In last twelve months or so we've seen a fundamental change in how telecoms are looking at Cloud computing. SPs are now taking and applying a lot of those datacenter and Cloud computing capabilities to offer new services to their customers, whether they are consumers, enterprises, or [small and medium businesses]. And as a result, the telecoms are looking for an entry level solution for service delivery from these datacenters - how do I provide Cloud services? – whether they are infrastructure as a service, video services, unified communications, collaboration services, and how can I deliver those to multiple different devices – TV, PC, mobile phones – simultaneously.

What we've done to address this opportunity is to put in place an SP class architecture that brings together a lot of these capabilities in the datacenter to be able to deliver a very customized set of experiences very rapidly. We then combine that set of data center capabilities with some of the next-generation IP-networking capabilities, ensuring that SPs are able to deliver the service all the way to the end user, on any end device.

If you look at the opportunities Cloud computing presents to you, it's all about being able to virtualize and deliver high efficiency whether you are offering applications, or services, or content, but you must also be able to ensure the delivery of the service: ensure the quality, make sure that it is secure, prioritizing the right applications to the end user. You've got to make sure that if somebody wants to watch a video stream, then that stream is arriving on the device quickly enough for consumption, that the picture is perfect, the quality is perfect, and if there happens to be an application running in the background or two or three other applications running on the same connection, that they don't interrupt and ruin that premium experience for the end customer.

A lot of what we've done is architecting an end-to-end solution for very intelligent, high-performance yet flexible delivery of services out of datacenters, using Cloud computing, all the way to the end customer

THORSTEN: These are all very valid points especially when you orchestrate a lot of different applications that reside somewhere in the Cloud. Application acceleration and reducing latency is definitely part of the problem, as is security. But what is Cisco's expertise there? Most of my clients are seeing Cisco a router box company, creating a connectivity solution, not necessarily a computing solution.

SIMON: In 2009 Cisco added to its solution set by launching a solution called "Unified Computing System". This is an integrated compute, management, and networking solution for datacenters and forms the foundation for Cloud services, effectually rewriting the rules of datacenter and Cloud execution by integrating a lot of the elements together. This was Cisco's entry into the computing space, but obviously building on a very strong networking base. We already provide a large proportion of the networking capabilities in both datacenters and across the telecoms industry. When you think about the implications of datacenters and Cloud computing the benefits all come from virtualization – the ability to virtualize the applications, the service, the content – and the ability to move [the application] around very flexibly so you can turn up the compute capacities, turn up the Cloud capacities, for whatever the current demand is. During daytime the datacenter may deliver a video, during nighttime it may be enterprise data processing; it may be to execute a follow-the-sun or follow-the-moon model. The core ability is to virtualize the service and the content. But you can only really do that if you virtualize the computing, storage and networking to be able to move that virtual content around, and also be able to deliver it to the end customer.

So our belief is that we've got a very unique ability to virtualize Cloud computing and networking and make it very simple to manage. It's an understatement to say that next-generation networking is just simply carriage of information. The ability to prioritize a service, the ability to identify these services, qualifying video higher than voice higher than data, the ability to make intelligent choices and decisions between subscribers or users or applications... it's going to be critical as the growth that we see now gradually overloads the existing capacity available to operators.

THORSTEN: So, how do I start as a carrier? I've my old services and my old ecosystem, and you come with a really interesting solution – what's the first step I need to do as a carrier?

SIMON: We are working with a number of carriers worldwide on exactly these questions at the moment. The natural evolution is to begin to trial and experiment with new applications and new services. A number of carriers are now looking at trialing Cloud computing services, or Cloud services for customers. The type of solution we're talking about should be integrated alongside existing datacenters, alongside existing services and activities. You can trial markets, put up a Cloud version of a service and see how many users adopt it and to what degree people want to move to that form of consumption from traditional consumption.

If I may illustrate that with a couple of ideas: the unified communications model is shifting traditional voice delivery towards voice-over-IP, and then combining voice-over-IP with location, presence, messaging, social media and mix/mash all those applications together. Those kinds of capabilities are really best delivered through a software-based, Cloud-based model. But a whole set of customers won't adopt that initially, so you probably want to offer both versions of a service and then tailor options and tailor growth, depending on what level of uptake you get.

There's also a second aspect that many of the telecoms are now experimenting with which is traditionally those services were painful or lengthy to install. You get them up and running and you leave them in place and never touch them again. The ability of Cloud services to virtually mount services, suddenly turn on an image and carry it over a virtual infrastructure from a virtual datacenter, enables you to very rapidly deliver new services, cut the delivery time from months down to minutes, and also it enables services that you previously never could produce. E.g. [Small and medium businesses] traditionally don't have datacenters, they don't have IT groups, they don't even have IT support. But now you can take all of these enterprise-class IP capabilities, put them up into a datacenter at the telecom, and telecoms can offer them through a web-based interface to [small and medium business] customers, and give them an enterprise-class, world-class IT Infrastructure.

Cloud also enables new business models: charging by the seat, by the month, bundling services together, turn them on, manage them, operate them, all remotely using web interfaces. So we're talking about some real, fundamental changes to the way users consume things.

THORSTEN: I can imagine that these changes will have an impact on the cost structure as well as the revenue structure. I can also implement business models now that were not possible before. And how is that going to change my organizational structure?

SIMON: In traditional telecoms you'd have a CTO role, and you'd have a CIO role, and you'd see different organizations under each group. We're now seeing a number of telecoms combine the CTO/CIO role. When you think about the evolution in architectural solutions that makes sense: increasingly you have a virtualized datacenter, and you're going to have virtualized computer applications spread throughout the network: content delivery networks, local content caching and delivery, even some of the core functionality of the telecoms and business itself.

Ultimately, you're going to have a fully virtualized infrastructure across the network as a whole, combined with the datacenter. Virtualized services and infrastructure shared between internal purposes (running it as a business for IT purposes) and external purposes (delivery of end customer services) will be a natural way to evolve.

The second part of your question is really around the business model and how Cloud is going to be consumed. We're seeing a very heavy shift towards on-demand consumption, initiated through web interfaces directly by a customer: Do-It-Yourself style models; pay by the day, by the minute, by the month, by the capacity, or by the compute power. So increasingly operators have to adapt to this more short-term, on-demand model which is leading them to the Cloud and the virtualization space. And we see a similar change in the way operators are looking to build that network: they often are looking more for assistance with build, operate, manage and transfer models; and the ability to be able to turn on features and capacities as users demand them.

THORSTEN: What does that mean for a telecom's infrastructure refresh rate? One of the concerns of procurement is always what kind of infrastructure refresh I have to do, or wiring refresh I have to do. With on-demand and virtualization – while flexible – it seems that erratic demand and different application, service, content, and network lifecycles could have a serious cost and management impact.

SIMON: The underlying hardware refresh rate remains the same, as the fundamental components remain the same. However the ability to reuse and repurpose equipment fundamentally changes with this new model. With a fully virtualized compute platform, and a fully virtualized network platform, it's very easy to re-provision, change the purpose of hardware, even virtually shift it within minutes between different roles and functions. This allows equipment to be reused and reduce purchasing requirements. Also instead of having to plan for the peak level of demand, you can actually plan around a fully virtualized distributed average demand. We believe that people will greatly improve their effectiveness and efficiency through virtualization. We investigated internally some of the financial returns that come from virtualization with Cisco's solution and changing the service delivery mode. It's highlighted between 20 to 80 percent reduction in [operational expenses], 20 to 80 percent reduction in the [capital expenses] required, and significant reduction in time to launch new services. There are some very compelling economic reasons why people will adopt virtualized data centers and Cloud services.

THORSTEN: That sounds like a massive reduction. How did you create your modeling and how did you calculate these numbers?

SIMON: You put side by side the traditional datacenter design, the operational processes, implementation processes, the number of applications, the number of services, the compute power necessary, the power, the cooling required. And then you put alongside that a fully virtualized model with a fully unified service delivery solution and virtualized network fabric behind that. The savings and improvements are then clearly highlighted. The full version of the tool is on our website at www.cisco.com/go/usd. The reductions come from less equipment, less power, less cooling and significant reductions in management and operations.

Some of the key benefits from the Cisco solution include the unified computing system that allows simplified management of computing and very efficient use of data center assets. We also have a very advanced and highly intelligent networking approach, unified fabric, which allows you to deliver a single unified fabric where any single cable can be logically used for a number of different purposes: to manage [local area network] traffic, to manage storage traffic and to manage the management traffic within a datacenter. A physical wire can actually be used for multiple different purposes and can be logically moved around to support Cloud.

When you start to run a unified fabric it can deliver a 70% percent reduction in the number of cables in the data center, the unified computing system delivers a significant reduction in the number of dedicated computing platforms, and a significant reduction in footprint, power requirements, port requirements, combined with the flexibility that virtual operations and web operations enable. Together – you reach the kind of numbers I was just talking about. And I think the industry itself is just beginning to internalize the potential that this represents. Not many people yet have made that linkage in terms of how much time and effort they can safe. Which is why we project it to be a very high growth area going forward.

THORSTEN: Another big problem is also compatibility with other Cloud services. From what you just explained it seems great for SP internal and external services, but in the future I expect that we are going to see many different type of clouds, there is not going to be just one cloud. With private and public clouds, hybrid clouds, how are we going to make sure that these things are all going to be compatible? On what things is Cisco working?

SIMON: Cisco today has a full set of services and solutions for privately built clouds (DC 3.0) – the ability for enterprises to build and execute for their own internal IT purposes. Cisco has an architecture (unified service delivery) and solutions for public Clouds, where service providers and telecoms put in place, launch, and deliver services externally to their end customers, and we're continuing to offering new elements and features and services to support this high growth market.

Most recently we've begun to announce the key capabilities both on the datacenter and network side to expand support for Cloud computing: web APIs, management interfaces to enable applications and services to be better informed about what resources are available, and be able to deploy and enable themselves much more rapidly. The intention of these APIs and capabilities is to enable users to interlink clouds together, the ability to connect private clouds and public clouds – as well as the ability to connect two public clouds together, so that service providers can enhance and share their services, or enable users to roam from place to place. Ultimately, what we'll see in the industry is there will be a set of standards being developed for inter-Cloud services and confederation between clouds, enabling service providers and enterprises to connect their clouds together and share their workloads. Cisco is doing a lot of work on these requirements to help accelerate that development, put in place some of the platforms, and set in place some of the elements to make that happen. That will be certainly an enabling factor in terms of open specifications and delivering new Cloud services. Because the ability to virtually move a workload from an enterprise to an external service provider and then back again is one of those critical capabilities in next generation Cloud services. We've already shown a couple of demos that show live working virtual machines can pooled in one datacenter, moved [virtually] a couple of hundred kilometers across to another datacenter, with the applications and services still running on the virtual machine. That kind of capability gives you real flexibility across geographies and across different markets as well.

THORSTEN: Were there any moments where you said "Oh my, I wouldn't have expected that!"?

SIMON: This is a new space, so during development a lot of new problems are occurring. If I think back maybe 18 months or so, probably one of the major realizations was how critical it was to link datacenters with networks. When you start to look at the implications of Cloud services, if you want to deliver a Cloud service where information is stored remotely and is accessible wherever, you really have got to be able to assure the delivery of that service. For the user that's got to be transparent, it has to be unimportant as where the information is, as long as you can get access to it. The implication of that is actually quite a lot of very explicit linkages between networks and datacenters to be able to make sure the service is delivered to the end customers, not just into and out of the datacenter. A second problem area was where there are a lot of linkages to be solved in the management of the datacenter structure, network elements, middleware, service orchestration, application interfaces, service delivery platforms, and application device-aware capabilities. There is a very complex issue just in the parts that need to be put together just to be able to make that really work well. That is quite an area where I would say that that's not well solved or standardized quite yet.

THORSTEN: What would happen if my solution wouldn't be device-aware? If I would just operate a datacenter with SLAs, not caring about the end devices?

SIMON: The challenge that comes up is delivery of services and your customer experience... well let's take a couple of examples. If you take consumer video, for example: if you as a service provider are providing premium video content to your customer - the latest TV series, the latest TV show, maybe broadcasting the Olympics, and I pay a premium for that service – and I request that service, I'm probably not going to be the only one doing so. There's going to be a whole lot of people who want to watch that Olympic event at the same time. All requests have slightly different times, so they look like individual requests for a lot of video content. And the risk is: it's a very popular event, so very heavy activity. Suddenly the network gets a lot of requests, the datacenter gets lots of requests, and ultimately the user experience can be very poor. Because the networks and the datacenter need to be aware that the user has a device with a small touch screen, so I can repurpose the content down to that device and size saving space, and because it is premium content therefore I need to prioritize it, over and above the rest of network traffic, this is the only way the user will get a good experience. It's that combination of being able to ensure a user experience while not relying on a best-efforts network, which would require enormous over-capacity to be built everywhere to be able to support that.

THORSTEN: So you are saying without that device-awareness I would have to massively overbuild my network.

SIMON: Yes, and you'd also run into some physical limitations, the amount of information space a telecom has available, for example, or the decision on how to proportion broadband networks that are going to be shared across multiple people.

THORSTEN: But is that a special problem of Cloud? Isn't that like a general problem you have?

SIMON: The problem is amplified by Cloud because the implication of cloud is that critical services and applications might be held entirely remotely. My willingness to make a transfer to a cloud service is really dependent on me always being able to get access and execute that cloud service as if it is really on whatever local device I have. The belief in cloud services will be significantly hit if the first few times people are trying it it's unprofessional or a poor experience.

THORSTEN: Very good point, you will harm your brand and your churn rate goes up. Probably the same thing that we see right now with the iPhones [on AT&T's network]: the first kind of device that has this kind of simple access to new application, a rich user interface, paired with entertainment and social networking capabilities.

SIMON: The more you give people simple interfaces, the more you give people all-you-can-eat packages, people will want more of this multimedia experience and you have got to be able to enable that for the whole population, not just a small proportion of early adopters.

THORSTEN: It's always interesting when you talk to a web-based startup that provides a VoIP or mobile video service, and they say boast about their thousands or ten million signed-up users. Just think about the 105 million live and concurrent users on mobile network right now, 24hrs a day.

SIMON: The interesting thing on the mobile networks is the rapidly growing demand for mobile connectivity: smartphones, mobile enabled PCs, electricity meters, logistics getting mobile enabled, etc. The volume of traffic is going to go up and up. and the congestion of services is going to increase. Increasingly intelligence and delivery of these services are going to be very important.

THORSTEN: What's the one thing that you would wish telecom telecoms would provide you, Cisco?

SIMON: I'm not sure if there's a specific Cisco requirement here, but I would say as an industry there are a couple of real opportunities: I think from a Cloud computing and Cloud services point of view the opportunity for the telecom industry is to get together and define a set of APIs for Cloud services. The opportunity for the content and media industry is to get together and offer a combined set of services and content and delivery over telecom networks. Telecoms also have the opportunity to reduce user complexity, if they take a lead from the iPhone, the

Nexus, the Blackberry-type experiences and put a very simple user interface and a very easy to execute model around all the complex services for a user they can deliver some incredible things here and hide the complexity.

I would love to see a significant change in the way services and interfaces could evolve beyond those offered today. There are millions of interesting Internet Cloud services out there. It's very hard to find them, and it's very hard to use them, and we still need that bridge to be able to get to them: a real simple interface and a simple and convenient way of finding them, I think that's a big opportunity for the telecom industry as this is classically what they do – engineer, market and simplify service complexity for the end user and cover the operational and bold challenges that go with that.

THORSTEN: Right: I pick up the phone and I have a dial-tone. And I can have a conversation across several different carriers.

SIMON: There is an interesting parallel on the mobile side as well. If you're an application developer for mobile devices you have to write 200, 300 different versions for every different handset type. Those are the levels of complexity SP should be able to master and simplify.

THORSTEN: Well, you could create an HTML5 interface and then have most of the [application] intelligence back in the Cloud. User interaction times on mobile devices are difficult: a half-second latency time is just too long for any rich and rapid user interface interaction. Of course I could use something like Mobile Sorcery or JavaFX or ISGi as an intermediate execution layer. But I'm skeptical that suddenly one C# development can reliably and simply compile for all kind of devices with all kinds of capabilities – that wasn't quite the experience I had in the past.

SIMON: I agree with you. But in terms of an aspirational goal, wouldn't it be a great opportunity to mix the innovation that we're seeing in the web services space and the Cloud services space and offer it over a very simple platform delivered to a mass market audience that telecoms represent?

THORSTEN: I have hope that there will be more interaction in the industry. Thank you for taking the time and sharing your thoughts and insights with us.

SIMON: My pleasure, thank you for the interesting conversation.

cloud
government
services
infrastructure
data
technology
customer
agencies
innovation
service
capabilities
technologies
business
federal
opportunity
sector
build
help
see
call
new
think
one
time
question
standards
applications
vision
development
environment
start
tremendous
experience
application
management
commercially
market
security
growth
role
need
essentially
priorities
Computing
products
make
taking
going
way
first
go
also
access
engage
opportunities
migrate
bring
two
get
ways
agency
participate
industry
introduction
procure
like
system
smaller
interoperability
Cloud
improve
research
Internet
available
play
economic
take
wave
right
corporate
believe
providers
product
verticals
strategy
competition
price
term
process
story
toward
big
larger
others
example
key
large
use
enterprise
experimenting
policies
using
Now
important
core
literally
sure
years
look
place
single
general
forms
TV
scale
economies
Number
sort
requirements
approach
idea
billion
complex
areas
public
might
spend
computing
dramatically
exciting
advantage
authentication
forward
prices
weeks
actually
long
may
design
national
move
every
around
private
shift
offer
domains
AppExchange
basically
seen

Federal Government, United States of America
February 16th, 2010

Change. If there is anything that really changed with the beginning of the Obama era – besides major healthcare reforms – it would be the use, ubiquity, and freshness of Obama's stance on Federal information and communication technology. For the first time in US history does America have a dedicated Federal CTO and CIO. Obama is profoundly convinced that the Federal Government has the duty to be a leader in pioneering the use of new technologies that are more efficient and economical. After all, it is the world's largest consumer of information technology and steward of taxpayer dollars. Ground-breaking initiatives by CIO Vivek Kendra and CTO Aneesh Chopra include the Open Government initiative, an Government App store, and the general move of Government applications and services into the Cloud.

In February 2010 Detecon's Daniel Kellmereit spoke with Federal CTO Aneesh Chopra. Directly coming from a briefing with President Obama in the Oval Office, relaxed but enthusiastic Aneesh – the "Indian George Clooney", as Jon Stewart put it – met Daniel in the local coffee shop right next to the White House to discuss the change and future of Federal Government.

DANIEL: Thanks for your time, Aneesh. So, let's get right to it – and talk about Cloud Computing. We have seen the wording change every couple of years; first we had Application Service Providing (ASP), then Software as a Service (SaaS), later On-demand, and now we call it Cloud. What is your vision of the Cloud and where do you think it is going to take us?

ANEESH: Well, let me begin with the larger vision. The American economy had experienced tremendous growth on what I would call the first wave of Internet based businesses, and the question mark in our economic team is to what extent we might see a second round of hyper-growth associated with the Web. We believe that Cloud Computing may offer that next wave of economic growth with businesses and startups flourishing alongside larger firms entering the cloud, and spurring a new wave of job creation and economic prosperity born out of the technology sector. So at the highest levels, we see this as a component of our overall economic growth strategy.

Number two, we also acknowledge that this is an opportunity for federal government leadership. Often times the federal government is a laggard by a decade or more in the adoption of technologies. Yet with a $76 billion information technology portfolio we have the ability to in some ways shape or form and participate in the early deployment of cloud services. So as the nature of the industry matures, we see our role as helping to migrate larger corporate enterprises toward the cloud – if we can – in a spirit of cooperation and collaboration.

Third, we actually believe that cloud computing could dramatically improve agency customer service, and for the lack of a better term new product development. The traditional lifecycle of a government IT project, from the time you conceive of an idea, Congress approves the idea and funds the idea, to execution, is measured in years. If we thoughtfully engage in the cloud, one could turnaround an idea in weeks. So the cycle-time for innovation to build new products and services that help to achieve national priorities and public mission, we think can be dramatically accelerated by virtue of taking advantage of cloud services.

So there is an economic growth story, there is a thoughtful introduction of efficiency into the federal government, and there is taking advantage of the capabilities of the cloud to roll out better customer services capabilities and improve our services for our national priorities.

DANIEL: So what role does the government play? Is it more of a laid back wait and see approach, or are you really playing and active role in standardizing and driving this ecosystem?

ANEESH: We have three roles, and I am responsible for two out the three. First, this is a nascent industry and there continues to be a need for R&D to build out some of the longer term capabilities that have to be put in place. The National Science Foundation has dedicated research and development funds to collaborate with the private sector to build that next generation infrastructure. That is one of the areas I am responsible for. We call that the Networking and Information Technology R&D Coordinating Agenda. We have a little over $3.5 billion in research and development. A portion of that will go toward Cloud Computing – already has – so that we have a long term roadmap. We recently announced collaborations with Microsoft, Google and others, shared research opportunities in that R&D framework.

Second, we have increased the NIST budget. NIST is the National Institute for Standards and Technology. They govern our standards work, and we provided an additional $70 million to think about standards opportunities for us, that essentially blend between our requirements and what is commercially available. For example, we know right of the bat, that for us to procure cloud services, we have to address two very important principles. First, what are the standards for interoperability: so that if we were to place our data in a cloud provider, how do we have an exit strategy to move that data, to avoid vendor lock-in. The standards in that regards I think are there but in need of further evaluation and work. So we will look at ways at which we can bring the private sector to the table and engage on the strategies to ensure interoperability.

Number two: we also acknowledge that the key security threat on the cloud is a security threat that cuts almost all web applications. How can you ensure user authentication, identity, call it whatever you what, so you avoid the cataclysmic effect of having all of your data corrupted and accessed? So on standards for user authentication, and on standards for interoperability, we intend to engage collaboratively with the private sector on ways in which we can move these agendas forward.

And last but not least, our CIO, responsible for the $76 billion in IT budget management if you will, is evaluating the right way to shift our resources from lesser performing assets, to building new data centers, continuing our $19-plus billion of infrastructure spend, and migrating our federal agencies toward a more shared infrastructure environment. Whether we call that private cloud, public cloud, public-private cloud, there may be various sets of operational elements here that we as customer will help to shape the market for. So it all blends together: long term research, collaborating on standards, and then acting with our own resources so that we make the right investments to participate.

DANIEL: Will citizens be able to participate in that cloud? Will they collaborate, will they input data, will they work together with whatever infrastructure we have? Or will this be or a closed, highly secure environment?

ANEESH: There are certain aspects where we are comfortable using existing commercially available cloud services. I come from experience as Virginia's Secretary of Technology: we deployed a solution in the cloud to simplify all the forms that you have to fill out to start a new business. In Virginia you literally have to go to eight or nine agencies to complete appropriate permits and paperwork, so rather than collapse all those government agencies and spend hundreds of millions of dollars trying to build a single data system, we simply used Salesforce.com. The development took six weeks. We had an intern that put all the forms on a conference room table, counting up all the times you ask for names, addresses and telephone numbers, repeated across the forms. We then built an intelligent form on Salesforce, where you could ask the question once and then raise additional questions as necessary, such as "are you intending to sell liquor", if yes, call up the session associated with the liquor permitting process. Such that at the end of the experience, the customer, which in this case is the entrepreneur, pushes the print button, and all the forms he would need are automatically prepared, using the data he entered once into the system.

Twenty-thousand businesses have used this application. Six weeks of development time and less than $150,000 have been spent to keep the capacity up. Sure, it does not have the full backend data integration, and information is not automatically entered, but it simplifies the customer experience, because they have now a single place to go. Entrepreneurs when surveyed told us we saved them on average three business days of search time, and the anxiety associated with it: do I have all forms I need, am I missing something, am I going to get in trouble? So you will see examples like that, where such data is at the voluntary use of the customer and is saved in the cloud.

DANIEL: Will the government cloud, let's call it that, ever move into very complex application domains? What we observe at the moment is that cloud applications focus mostly on highly standardized process domains like CRM and HR management, as well as less complex application verticals.

ANEESH: Well, the first sort of commercialized cloud services in the federal government is NASA's Nebula. NASA has a tremendous amount of modeling and simulation capabilities, and other data capabilities that one would consider as on the high end of computing. That is a government cloud, meaning we are not using a commercially available service, where the applications as you are describing them are more productized. We have already seen that. The Department of Defense has an IT infrastructure program, where they have been running cloud applications in the defense infrastructure.

So we are starting on both extremes, there is www.apps.gov, where we have access to a series of commercially available services, where you simply have the light weight apps, often for social engagement purposes. Then on the other extreme you have government clouds where we are experimenting with the Defense Department and NASA Nebula, and the question mark is how you start to bring it in between.

Frankly that is the excitement – when we see IBM, Google, Microsoft and others all start to say they will move their enterprise class software into the cloud it opens up a much more interesting dialogue around how we procure their services moving forward.

DANIEL: That is great to hear. To be honest, I personally haven't seen many applications with complex underlying processes moving into a cloud environment. In this field a lot of software remains highly customized and in "closed" corporate datacenters. Of course these companies use virtualization technologies to improve portability, manageability and compatibility of applications, but this is only a small piece of the complete cloud story and vision.

So how have business, technology and our economy changed since the introduction of cloud technologies? What are the core shifts you are observing?

ANEESH: The core shift is that the distance between business strategy and concept to prototyping version 1.0 of technology has collapsed 80 – 90% in productized verticals due to cloud technologies. Let's take the example of a typical business executive who is buying a business intelligence product. He has to set the requirements with his CIO's office. He has to run a procurement process. Then he has to build the infrastructure – what are the servers that we need, what are the security protocols, what data will we be analyzing, how will people access that data – all of that before you have even started the design of the actual customer experience. That process takes months, charitably, years realistically. Now by taking advantage of cloud service you are essentially renting that infrastructure, including the concepts of what you want to analyze and how it translates into your operating units. This literally could happen in weeks, if not months.

So what does that mean? I believe that means we will see a dramatically accelerated role for technology in new product development, in customer experience and design, and in other areas that require a more agile footprint. The probability that your enterprise class financial system is going to migrate to the cloud tomorrow – not my priority – but the fact that you can adjust services and products, the way you engage your external stake holders, absolutely will be easier and faster through the provisioning of cloud. Intellectually we are making the leap that happened in the electricity business however many decades ago, where you might historically have to take care of your own energy usage and you have to have an energy management staff person to make sure you have the power you need. Now you have a contract with the utility company and you are basically plugged in and it works. We want to get to a vision where in technology, the application of technology, is as easy as plugging it in and it works.

DANIEL: It's obvious that US technology companies, and also the US government, are taking a leadership position in Cloud Computing. We are working with telecoms and other infrastructure providers, advising clients on strategy and technology related topics. How can they learn from your initiatives and how can they make sure to make the right moves at the right time?

ANEESH: I'm a disciple of a professor by the name of C.K. Prahalad who authored the book "The Fortune at the Bottom of the Pyramid". His fundamental premise is that in a globalized marketplace, engineering products and services that meet the needs of emerging economies, frankly the world's poor, have the dual benefit of introducing new concepts one can re-import into domestic markets, but also allow for a new channel of profitability.

An article in Harvard Business Review by Jeffrey Immelt highlighted GE's approach in this regard. They call it reverse innovation, and it really speaks to the translation of Prahalad's vision into a corporate environment. Now why do I bring that up? GE has introduced medical diagnostic equipment at 85% of the price point, which is profitable in rural China. If we re-imported that technology and innovation into the US domestic market to expand the marketplace – redeployed to the US with more modular design, more sharable intellectual property. So I actually see tremendous opportunities looking at emerging economies as target markets for US firms wishing to exploit principles of reverse innovation. Innovate in these countries where price constraints are such that traditional services are too costly. Gather that feedback and learning and re-import that feedback as GE describes. I think that same play will be played out in the US technology sector and if we don't play it, others will. And that is why I think it is an opportunity for revenue growth and an opportunity for the next wave of success in the technology sector that fosters job growth.

DANIEL: When looking at reverse innovation, how important are economies of scale? How important is it to be a large nation to be successful in this business? Do you also see opportunities for smaller countries to play a role in that game – or do you have to be China or India or some other larger country to be of relevance?

ANEESH: It is not scale at the government level. It is scale at the infrastructure application level that drives the cloud. So if the federal government were to "en mass" utilize a particular cloud service, you can imagine a great deal of industrial strength and security and other capabilities will have to be introduced as part of that mechanism.

By the way, we are experimenting with that. In short order – by the time the book will appear – we'll have taken an ideas management system in the cloud which had been utilized by federal agencies, to launch. This is a one-off modest implementation where the general service administration on behalf of the 24 largest federal agencies essentially would offer a single instance of this cloud solution to gather public input. In just 60 days we will have taken this service to launch, so is an example that it should not matter where you start.

A big government may start in a simple small agency, not unlike smaller governments around the world. You don't have to be a large enterprise. Because in fact of what I said earlier about the price constraints: it actually may be that you are more willing to tolerate risk in a smaller government to experiment with these services and have that be the alpha or beta site for some of these activities. That could then feed the confidence measures that you would need to have it scaled in the larger economies.

DANIEL: Personally, what application domains are you most passionate about in cloud computing? What are the verticals that will see the biggest impact through introduction of cloud technologies?

ANEESH: Two areas that are obvious: customer relationship management is sort of the first foray into the cloud and you can imagine on a dramatic scale how that can fundamentally shift your relationship with the government. We are experimenting with strategies to see how we can improve customer service by having a thoughtful way of engaging individuals. So our usage of that, which is a mature service in the cloud, will be exciting to watch in Washington.

But for the industry as a whole, analytics in the cloud I think is a tremendous opportunity, because more and more of us are gathering data at a rapid clip. The cell phone I have in my pocket emits location data every minute or so to the wireless towers. If you wanted a digital footprint of everywhere I have been you could literally look at the terabytes of data that carriers might have on our patterns. They could then proceed to analyze that data and think about new and creative ways to monetize, without compromising privacy. This is a wonderful opportunity to understand how cloud providers could help take large data sets and help provide immediate value. So the role of big data, the role of analytics, data driven decision making, I think that is a key opportunity for technology in general.

And there is no reason not to offer these services in the cloud. Why do I say this? One of the key features of the cloud is that it allows for the share-ability of intellectual property on top of base platforms. Within Salesforce.com as an example you have the AppExchange. Rapid product improvement cycles by sharing components that are in the AppExchange, help you build what you need. In fact that's how we did the Virginia story called the "Base Salesforce Application". And we added 2 apps that ran in the AppExchange to compliment it. The e-forms and some basic rules-engine workflows plus the core CRM combined gave us a six week delivery cycle to get version one out of the door.

DANIEL: If you look at consumer facing technologies, what are the devices and services that will drive cloud adoption from a user-centric perspective? Is it Next Generation TV, or new touch screen devices and tablets?

ANEESH: Internet-enabled TV is a technology that has been demonstrated at the CES for a while but has finally matured. The notion that I can have a Skype call to my doctor might democratize access to telemedicine, distance learning and a variety of other national priorities.

The Department of Veterans Affairs today spends on average $1,000 to $1,200 to equip a patient with remote monitoring and telemedicine capability so that they can help to keep you healthy after you have been discharged from the hospital. And they have published their results – a 19% reduction in re-admission rates of patients, and an improvement in quality and improvements of costs. If you can have Internet enabled TV where video conferencing providers like Skype can

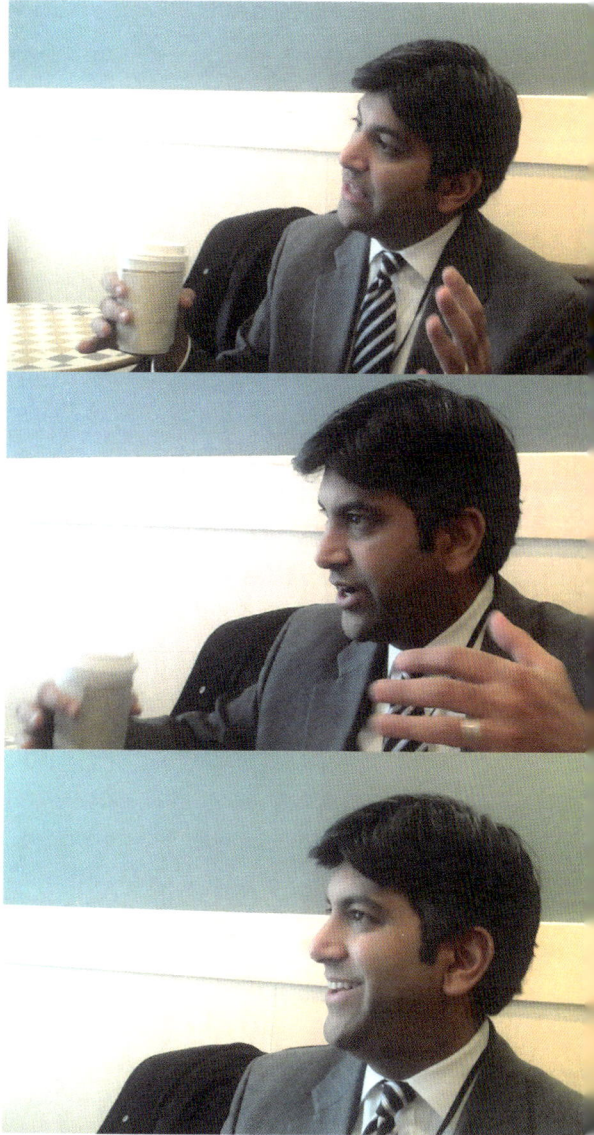

essentially cut that $1,200 capital spend by another 90%, all of a sudden every patient, when they are discharged from the hospital, will have the ability to communicate back with their physician or nurse practitioner to make sure that they are monitored on a frequent basis in the convenience of their own home. Much more so than 3D TV and some of those sort of exciting components, I'm particularly motivated by the commercialization of these very lightweight Internet connection points that could advance national priorities.

DANIEL: One last question, before we close off. When I look at the US government and innovation in general, I can see a contradiction where the government procures somewhat out-dated technologies one the one hand, but tries to drive the very front-end of technology innovation on the other. How will that ever align and build a common and orchestrated strategy that makes sense?

ANEESH: Three parts. Part number one is on the leading edge side. We have issued a policy called the Open Government Directory in December. That has directed agencies to basically rethink their approach to the market. One of the requirements is the delivery of a new set of policies around prices and competition. So on the cutting edge you can use prices and competition policies to bring in cutting edge cloud services. This will be a focus area when we issue the prices and competition policies.

Number 2: there are certain centralized platforms that we could procure, that will help to democratize agency access to services so that they don't have to go through a long procurement cycle. They can go essentially through it once and then others can plug into that infrastructure. We are going to be pushing in that regard.

Number 3: by basically holding the agencies accountable through rating their infrastructure spend, we will create market pressure within the agencies to shift to requirements that would include cloud service providers as part of their mix, so that they can achieve their data center consolidation goals and their IT infrastructure efficiency goals. In this way we are creating the market conditions whereby the agencies will naturally migrate more and more of their capabilities into the cloud. But for some of those basic elements, the interoperability question and the online authentication question are two that will require further evaluation before this is going to be realized as a big vision.

DANIEL: A huge effort, and an exciting one to watch going forward. Thank you very much for taking the time for this interview.

ANEESH: My pleasure.

business

telecoms

think computing

Cloud going scale

services something network well

storage

services people think network

GoGrid

companies Cloud data

like provider really got one

brand need great marketplace

interesting perspective large much see want

automation Web name

commoditized connectivity different

infrastructure system units

relevant emerging play

compute telecom IP lot

things Amazon actual

Customers innovation

Telecoms now something order

done good balancing integrated installations center

decide enterprises partner technology equipment using looking

Infrastructure-as-a-Service transformational transit economy

two still take better largest access

labor pieces networks also already

quite thing next One back

addresses markets operations

cost automated either commodity

billing pricing example

supposedly persistent hardware CIOs

making delivering offers load

pundits many larger service

capacity offering rollouts card

far unit operating

clients difference features existing credit right

costs providers manage level

big SkySys elements ones yet

focus achieving particular generation etc

performance often new compared time may number

planned

GoGrid
March 3rd, 2010

Conveniently located in the San Francisco Telecom Center, just south of San Francisco's Financial District, is GoGrid's West Coast Datacenter. A Valley native, GoGrid is the retail brand of its mother company ServePath, Inc. and provides infrastructure hosting with extensive expertise and experience in running complex, on-demand clouds on dedicated and mixed server infrastructures. GoGrid's multi-tier cloud computing platform comes with a web as well as a REST-like Application Programming Interface. Customers continuously rave about the managed load balancing, low latency, and high availability.

In March 2010 Detecon's Thorsten Claus spoke with John Keagy, CEO of GoGrid. Leading an exuberant and enthusiastic team – the company is employee-owned, and yes, John really *is* Dr. GoGrid – John explained why he thinks that telecoms will get marginalized if they only concentrate on network build-out; and how new alliances, partnerships, and service offerings of telecoms could look like.

THORSTEN: How do you position GoGrid in the market?

JOHN: From the telecom provider perspective we are next to Amazon the largest Infrastructure-as-a-Service provider with a feature-complete offering. What I mean by that is that we are offering both Linux and Windows servers in the Cloud with integrated cloud storage, integrated load balancing, hourly billing, and we're doing it for nearly 10,000 customers in more than 100 countries.

THORSTEN: Are these customers mainly large companies or smaller companies?

JOHN: Those are Web 2.0 companies and Web 2.0 business units of large enterprises. Macy's with their iPhone app is a good example of a larger business that has a Web 2.0 business unit that's doing business with us.

THORSTEN: What's so attractive about your offerings that specifically attract Web 2.0 players or business units?

JOHN: These are people with production workloads and they like us because we are true to standard datacenter computing concepts such as persistent storage, persistent IP addresses, contiguous IP addresses – real reliance. Those are standards of protectoral constructs that they're using already in on-premises computing or in their other installations and that allow them to easily work with GoGrid.

THORSTEN: Are any of your clients, especially business units of larger companies, concerned that unanticipated capacity bursts could increase the costs massively?

JOHN: No. GoGrid offers very competitive pricing with great volume pricing plans that with scale make our pricing even more affordable than Amazon. And that's one of the great things about Cloud computing: you pay-as-you go. So cost control is much easier.

THORSTEN: Who is signing the check right now? Is that someone within the Web 2.0 business units ordering services and putting them on her credit card, or is that more like a C-level decision enabling business entities to order services?

JOHN: It's typical done at the business unit level, and it may or may not be done with a credit card, other times it's done by invoice.

THORSTEN: You are an infrastructure provider and not a network provider. Early virtualization of desktops and cloud computing didn't take off as planned because of challenges in the network. One would have thought that network providers would have a bigger play in Cloud because you need to transport data from A to B, do some load balancing, have many distributed data centers, etc.

JOHN: I think connectivity is something we do quite well at GoGrid. There have been a number of independent parties that have compared data transfer performance on our network against Amazon and other networks and we've done a better job. Not many enterprises have that scale to be able to have the connectivity that you need to deliver good global performance. But there's not much rocket science to do it.

THORSTEN: So what's the telecoms' play then in this space? How do you see telecoms positioning themselves?

JOHN: Telecoms have to decide if they want to be in the business of compute and storage. Network and access has been commoditized. Well, network costs are still making up a large part of the IT economy, but that's shrinking quite rapidly. Cloud computing and storage is so much analogous to networks because it's a scalable business that can be delivered on an automated basis where there can be some efficiencies had from scale.

THORSTEN: Is there any particular innovation you would like to see from telecom or other network access providers that you would love to see to either create a new business, grow or mange your business better, or enable a different kind of business model?

JOHN: No, because IP transit is really all we need. I think all the innovation that's going on is happening in compute and storage. Services that typically reside on transit circuits like voice have already been commoditized in IT. So: no, nothing I can think of. Everything's IP from my perspective.

THORSTEN: That's very interesting. Telecoms traditionally had large investments in network infrastructure rollouts. We're seeing next generation mobile and fixed networks being planned, 4G and fiber rollouts. Telecoms fear that they cannot survive on micro-margins in commoditized markets. Is there no opportunity or need for value-added connectivity services for Cloud? Can telecoms only play in storage and compute segments?

JOHN: It's the only thing I can think of. I'm sorry to say it so blandly but I think IT communications has been resolved to one commodity: IP transit. And the prices are dropping still.

THORSTEN: What's the next big step you thing the cloud industry is going to take? How are Software-as-a-Service, Platform-as-a-Service, or Infrastructure-as-a-Service markets going to develop?

JOHN: There is going to be a rush of new vendors into the marketplace. One of the most interesting things will be to see if any of the great brand names of telecom and computing really do anything relevant to what GoGrid and Amazon are doing. Right now you've got GoGrid, which is not a main brand, and you've got a company that's well known for selling books leading the charge in cloud computing. When are we going to have a great brand name in computing like HP do something that's relevant – relevant services at a relevant price point? When are we going to have a great brand name in telecommunications really establish itself as doing something relevant?

THORSTEN: If telecoms partner with you or help their clients using your services, what would be the single most interesting innovation they would look back on in ten years and say "We did the innovation X and that really made us shine again."? Where do you think are the great opportunities to engage with telecoms from your perspective?

JOHN: The great thing that's going on in IT is automation. Anything that's related to automation is what we will be looking back on as transformational. Since telecoms are well suited for delivering automated services at scale it makes a lot of sense for telecoms to participate in this.

THORSTEN: But isn't that your business or Amazon's business right now already, to delivery something at scale? Why would you need or want to partner with a telecom?

JOHN: It would be nice to have some validation in the marketplace. To have a big name brand really do it well would be good for the marketplace.

THORSTEN: Why is that?

JOHN: The marketplace in general is lacking a great brand name that's really doing infrastructure and service well. Until the marketplace has that there will be a lot of doubters.

THORSTEN: You mentioned that automation will be transformational. Aren't you afraid that your business is going to be completely automated and that you could be reduced to a commodity as well?

JOHN: There is that risk, but we own the automation technology. We're not a reseller of somebody else's automation technology; we're the ones that are developing the capability of automating IT. If you're looking at the 1.5 trillion dollar IT economy there's two huge pieces that I think are going to go away: One is telecom cost and

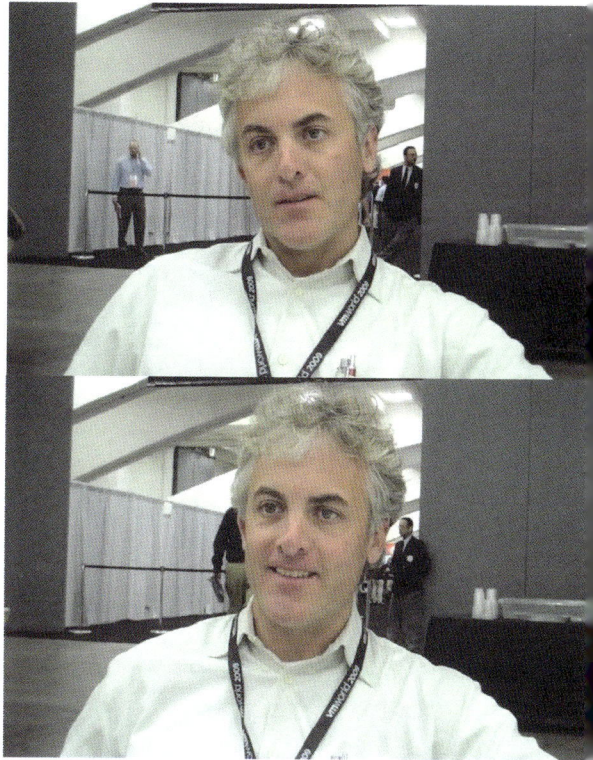

the other is labor. Those are two of the largest pieces of what people spend money on in the IT economy and I think that those are going to be driven towards zero.

THORSTEN: CIOs would argue that the labor costs stay the same but focus changes of their internal IT organization; people don't do so much plugging, installations, hardware, setup, and maintenance of actual boxes anymore but focus more on business planning, business processes, negotiations, SLAs, and contracting.

JOHN: Most certainly those are necessary things for businesses to do. But I also think that those are executive level skills, business decision making requirements that CIOs have regardless of what they're doing. Whether they provide IT in-house or whether they've outsourced it: they have to decide how they want manage their business and what capabilities and services they need. I'm not that's entirely relevant...

THORSTEN: What is the most interesting and unexpected lesson-learned during your operations?

JOHN: I'm every day amazed how different … There are a lot of pundits in Cloud computing that think they know what the marketplace needs. I compare what the pundits say versus what our actual customers and actual users say. And there's a vast difference between theory and reality.

JOHN: For example, Features of people supposedly want, what's supposedly important to customers and what's not. It's very different than what people like to blog about. There's a lengthy list of small items and some big architectural ones.

THORSTEN: Like Cloud-in-a-box – which more often than not turns out to be "in-a-box" but not "Cloud"?

JOHN: Sure! There are also a lot of folks out there trying to sell cloud operating systems to telecoms. But so far the only one that's proven to operate to a scale telecoms would need is our SkySys operating system. There's just nobody else that's achieving that scale, not even in order of magnitude. In terms of a cohesive system that offers all the elements of compute, storage, billing, and customer management and operations, in something that pundits don't write about but that is a very practical need, and that's the operational tool for delivering the service. That's something that all these other companies that think they've got a system that can distribute the ends on the hardware. They think they've got the system. That's one of 36 elements of our system.

THORSTEN: Is there a difference between mature and emerging telecoms and what SkySys allows them to do or how they would use it?

JOHN: So far we are working only with well established telecoms: SkySys is targeting your existing data center, your existing hardware. Emerging telecoms often either don't have that data center yet, or – exactly because they are emerging – they don't want to have them. Purchase equipment, "rack and stack" that equipment, and manage capacity is not yet a best practice for emerging telecoms. It could be.

THORSTEN: Thank you for sharing your time and insights.

architecture ago moving everybody two twenty announced ActiveSync scale lot enterprise like able new happen three connection partnerships software using just people Microsoft go things recently architectures now four including becoming computing devices help even iPhone commodity point better single mail surprised technology primarily Amazon getting running benefit Cloud Exchange platforms reason right less experience Blackberry company model business issues services much space small also systems People next looking one approach generation expect talk users browser case number system thousand access years especially products talking see telecoms way going get need meetings really well mobile Apps applications thing medium move costs email customers maybe online anymore think started first hard Enterprise critical employees cloud kind pretty rapidly currently large problem companies take something relationships send phase speed device Google service second time megabit start operational huge Server Initially seeing else Docs

Google
February 24th, 2010

With over 62 percent market share of search engine traffic in the US and over 85 percent globally, search giant Google is well known and almost wouldn't need any sorts of introduction. But Google is more than search. As a company, Google focuses on three key areas: Search, Ads and Apps. Google is not just leveraging Cloud for internal operational effectiveness and efficiency, but has also many public Cloud offerings.

In February 2010 Detecon's Thorsten Claus spoke with Raju Gulabani. Raju is Project Management Director for Google Apps and known in the Valley for his level-headed, calm, and visionary view as well as for his gentle and seemingly impeccable product launches. Thorsten talked with Raju about how enterprises move toward Cloud services, how a successful transition from legacy services can be done in large organizations, and what the future play of telecoms is.

THORSTEN: Why don't you start off with telling our readers a little bit what your position is with Google and what you are doing?

RAJU: I'm project management director for Google Apps. I started the Google Apps effort almost four years ago. For the first two years, I ran all development and product management. Now I'm responsible for the various aspects of it, including primarily the mobile applications that we use for Google Apps and enterprise customers.

THORSTEN: What was your starting point, and how did that change?

RAJU: Initially we were primarily focused on a wholesale computing service. That was in 2006. And we early decided to take these services from the consumer space into the enterprise and business space because we were getting a lot of customer requests for it. But as we got into it, it became obvious to us that one of the key requirements was the mobile access to the data, especially as the data moves to the Cloud. It gets a lot more easily accessible in the Cloud because you don't have to establish a Virtual Private Network connection to access the server, as is the case with many other mail systems and access to data. Once people start using our system from a browser they realize that they'll be able to access their data pretty much from any device – whether it's your corporate PC or a laptop or a home computer. And then you start to think about how to use it from a mobile device. Initially we looked at Blackberry devices, later more and more at iPhones, Android, and other devices – anything with a web browser. We just saw that mobile was a critical access point for data.

THORSTEN: What were some common issues with mobile data access and cloud computing you encountered?

RAJU: That's a good question. So there are a lot of things that were really hard initially to do.

The first one is that the initial architecture of all the mobile devices when they came was designed for client-server kind of applications. A Blackberry, which is the most used mobile device in the enterprise and for business users, the architecture requires the Blackberry Enterprise Server (BES) which sits behind the firewall in an enterprise and connects to a [Microsoft] Exchange server. When the iPhone became enterprise-enabled Apple really designed it to talk to [Microsoft] Exchange servers. They licensed the ActiveSync stack from Microsoft. Both of these device classes are really looking at currently existing architectures in order to sell these devices to the enterprise.

When you come in with a cloud computing architecture like we have with Google Apps, you have to take these devices and even though they believe they're talking to the old-generation architecture, they need to talk to the new generation cloud computing. We had to deal with a lot of interoperability issues, like service stability. In case of Blackberry, we have a solution called Google Apps for Blackberry Enterprise Server, which essentially allows a Blackberry to talk to a Blackberry Enterprise Server and the Blackberry Enterprise Server – instead of talking to [Microsoft] Exchange – talks to us.

We had to take the devices which were talking to previous generation architectures and bring them over to next generation cloud computing architectures. In case of the iPhone and other ActiveSync devices, we license the Microsoft protocols and implemented it in the Cloud so that an iPhone that was talking ActiveSync to a [Microsoft] Exchange server is now talking to our Cloud.

So we had to do a lot of things like that which were initially really hard to do.

THORSTEN: With Cloud I don't see your efforts anymore. Does that mean that people de-value suddenly tasks such as backup, maintenance, updates, or connector development and just expect them to happen for free?

RAJU: People don't expect that to happen for free, but the economic model is very different. Folks like Google are able to do it at a fraction of what you'll be able to do, even as a large company, because we're running at a scale with hundreds of millions of users. A large company has maybe two hundred thousand employees, maybe just fifty thousand. So your operational costs are a lot higher because you're not getting our benefit of scale.

THORSTEN: A while ago I was looking at the email system at a major global bank. On days when a lot of company reports were due they had a humongous data volume as most of the reports were sent to distribution lists – a single report was stored over and over again in everyone's inbox and often forwarded to even more recipients. When [Google] Wave was announced, everyone was really excited because you only have one single copy of any asset in the network – is that something you're looking into as well, something like a single data storage point for many copies?

RAJU: Because Google Apps consists of email as well as our Docs and Spreadsheet services, more and more people are not forwarding attachments. Instead they create a document or a spreadsheet or a presentation online and just send a link. All of a sudden all the data that is going through on your network is significantly lower. And even if you do attachments we don't necessarily send those over to your PC: they stay on the server and there is a link in the email to those documents. You can certainly download them. You can also convert them over to our Docs kind of a format and store it in the Cloud and see it in a browser.

All of these options essentially reduce the amount of traffic: With traditional mail services all this data was sucked in on your broadband connection. Whereas with the browser, that's just sitting on the server: you're seeing the currently needed pages in a browser, and the time to display the current page is a lot less – unless you choose to download it.

THORSTEN: Telecoms have a natural interest in controlling the amount of traffic, especially on the cellular side. Do you have any telecom partnerships?

RAJU: We have some partnerships, and I think you will see more in the future. Cloud computing is a relatively new phenomenon. We started four years ago and we were one of the first ones to start doing it at scale. It's just starting to get mainstream now. In 2009 we started to see very large companies hosting their data in the cloud. As typical with new services, they usually tend to go direct first because the initial customers will bear their business on a new technology. Once you sell to more and more mainstream customers that's when usually channels get developed. So I think you'll see for us a lot of partnering relationships with vendors of all kinds, including telecoms. Because a lot of telecoms have successfully been able to penetrate both large and small companies. Those relationships are very, very relevant. Telecoms are going to be natural partners for us, you will see that.

THORSTEN: What kind of services are we going to see next?

RAJU: First thing that you will see are horizontal commodity-like services. Email is an example of that. Traditionally you need to run that service yourself, and the only reason you do that is because that's traditionally how it's been done and there's not been a better choice. This is like going back to the old days: when you were a car manufacturer you had to know how to build a power plant next to your car manufacturing plant because that was pretty much the only way to drive things. Today you just connect to the grid and you get the power from it – well, theoretically… It's the same sort of idea: anything that's a horizontally, broadly used service will move to the Cloud pretty rapidly.

I expect that over the next three years company phone systems will move to the Cloud at large scale, and I think it will start with small and medium businesses. There is no reason for a phone system to sit on site at all, besides some benefit to the liability.

Anything that you currently get as a server from a vendor but that is not something where your own understanding of your core business is critical to running that yourself – it's not a custom piece of software you're writing – that's going to move to the Cloud.

THORSTEN: Connectivity becomes really important with all these commodity services moving to the Cloud. What would you wish telecoms would provide you with?

RAJU: The first thing is speed. I think that we get a couple of megabit per second in the US over cable modems and on DSL. If we could get that to 100 megabit per second, that would be huge. Google has been recently announced that we're doing Gigabit testing for a limited number of customers.

High speed connections become crucial, also for us at Google. We have 49 offices worldwide, and there is a culture of using video conferencing a lot, but that gets to be very expensive, getting conference rooms with hardware and so on and having all these connections. If you could do that kind of thing on a regular broadband Internet connection, that would be huge.

And there are other applications like online meetings where the kind of experience you get online is going to get more and more like what you get face-to-face, which sometimes is a huge time sink: For a 2h meeting we're not making a 2 or 3 day trip anymore. I think online meetings will be a new application for telecoms.

THORSTEN: But isn't high speed Internet connection rather a business problem than a technology problem? Fiber-to-the-Home is no rocket science, but doing fiber rollouts cost-effectively and creating a business model that support these costs is tricky for telecoms.

RAJU: It's a chicken-and-egg problem. People will want to be sure that they have a high-speed connection to their Cloud data and that this connection and Cloud data is available all the time. The three things are speed, availability, and reliability. Being able to access data from anywhere is part of that as well: Yes, we have 3G cellular wireless access, but that is still much too slow.

THORSTEN: Are we going to see more platforms in the Cloud where customers build their own applications, or do you rather see an emerging service industry: more and more services provided in the Cloud, less and less platforms?

RAJU: That's a good question. I think this is a two phased approach. The first phase was the move of critical applications to the Cloud. The second phase that is going to happen over the next couple of years is moving platforms into the Cloud. Google has Google App Engine, which is our platform in the Cloud. Amazon has Amazon Web Services (AWS). Microsoft has recently announced Azure. We are at the beginning stages of that.

THORSTEN: How is the market going to develop? Is it more about strategic partnerships? Is it going to be technology innovation driven? Or is it going to be dominated by the smartest business or economic model for platforms?

RAJU: It's hard to say which way it's going to shake out. The three major players in this space are going to be Google, Amazon, and Microsoft. Each of them is going to have strength and weaknesses within its space. I don't know what's going to be the winning formula right now. Business-wise, it's early. To some extend it's going to be the number of applications that are available, one way or another. It's also going to depend on the operational costs, and in case of Microsoft it also depends on the legacy Microsoft Windows platform.

THORSTEN: During your four years as a product manager for Google Apps, what was the lesson learned that surprised you the most?

RAJU: When we started out the conventional wisdom was that Cloud Computing was going to first happen in the small and medium business space. Most of us were pleasantly surprised to see how quickly the traction has come in the large enterprise space. We're seeing a lot of companies moving their entire fifteen to twenty thousand employees to the Cloud in one flip-of-the-switch, if you will, instead of saying "well, I'm going to first give it to people that don't have email services, maybe they're not knowledge workers, and then, I think, a phased approach for the next two divisions," and so on.

It pleasantly surprised us that Enterprise Cloud services are actually accepted much more rapidly.

THORSTEN: I had the same impression – It's not a small business anymore, it's really large corporations.

RAJU: That's right. For small and medium business it's a no-brainer: for them the value proposition is very clear. But for large enterprises people believed that they would want more control, they don't necessarily need to save money as much, privacy and security issues are much more significant, or that they have a lot of complexity, legacy hardware and software that has to be interfaced to, etc. Enterprise Cloud has clearly a lot more issues. We didn't expect us to be making as much progress there as we have made.

THORSTEN: What do you recommend to Enterprise clients with twenty thousand or more people – what should they be very careful or aware of?

RAJU: We recommend that it's actually better if they move everybody over at once.

THORSTEN: Why is that?

RAJU: It happens to be advantageous over being, say, half on Lotus Notes or Exchange: Whenever you have two systems that causes more complexity, and more issues.

We recommend moving people over it in two phases: Definitely try the service out in your environment with a limited number of users, a number which is large enough to give you a sense of what kind of issues you're going to run into. Typically at a ten to twenty thousand employee company start with a hundred employees and then go to a thousand over three months. And if that works out, then flip everybody over.

The second thing we do is when we move over these large customers we manage the deployment very carefully: The week that they move everybody over we send twenty or thirty "Googlers" to their offices from across the country. We then go over and spend a week being help desk, if you will, or Google hosts, at the company. Anyone that is running into any questions or any issues we're able to help them.

The other thing that we found is that a lot of people already know our products. The user experience is the same for our consumer products versus business products because 90% of the features are the same. People either already know our products or there are people in the office who do.

THORSTEN: As your services become ubiquitous, are you afraid of becoming a low-priced commodity?

RAJU: Well, I don't think we are becoming a commodity, but we are offering our services at a fairly low price, just because our operational costs are so low that we can do that. Gmail, in fact, is a very differentiated service: No one else does email the way we do it. For one, we have much better spam filtering than anybody else. Second, our conversations feature is unique. Third, the way we do search: I can find old emails within the 10 or 20 Gigabytes of emails that I'm currently using much more rapidly than I ever did with Outlook running on my laptop. Those sorts of things are not commodity, and that's our secret sauce.

THORSTEN: I really appreciate your insights into Google Apps, thank you very much for your time. I'm looking forward to seeing more telecom partnerships coming up.

RAJU: Agreed. Happy to have talked to you.

five even take day
back people new wi-fi users
pricing example ultimately
essentially department billing build give
find telecoms carriers difficult servers time make well
public clouds notion matter large right buy big
used going applications ago
kind today user corporation anymore relationship good cost
blackberry easy way now computing
mobile single per whatever multiple many use market bring
always reason
deutsche get think system capital
go like years much percent
first
center know services
cloud provide
case store value one actually
devices
believe solution application infrastructure
pay around answer bank mitsubishi
hardware business
lot really stuff
customers idea end need iphone
across centers multi-tenancy
expenditure deliver web thing
put using sell support
talking maybe access ntt
question device utility
expenditures customer
ownership software
app service different someone providers apps
interface want
version sales crm private
salesforce.com server also data especially
lawson operational
world bandwidth

Intalio

December 1ˢᵗ, 2009

Tucked away next to a micro-brewery and a coffee shop on a sunny and quiet side streets of Palo Alto, CA, lies the humble headquarter of one of the earliest Enterprise Cloud companies. Founded in 1999, Intalio has over 500 customers, across all vertical industries, in over 50 countries. Intalio is made up of two divisions, Intalio|Cloud and Intalio|Works. While Intalio|Cloud leverages a combination of hardware and software to deliver an enterprise-grade cloud computing experience with dynamic provisioning and elastic scalability, the Intalio|Works division develops and supports industry-leading applications and infrastructure. From the beginning Intalio was committed to create and support a sustainable Commercial Open Source Model – seven of the 56 top-level projects currently managed by the Apache Software Foundation were initiated by Intalio.

In early December 2009 Detecon's Thorsten Claus spoke with Ismael Ghalimi, founder and CEO of Intalio. A passionate entrepreneur and fervent industry observer, Ismael evangelizes his vision of Cloud as a founder of the Monolab|Workspace and advisor to several high-tech companies, including 3TERA, AdventNet (a.k.a. Zoho), EchoSign, and Egnyte.

THORSTEN: What does Intalio do?

ISMAEL: We provide a vertically integrated solution for private cloud computing. People are very intrigued by the notion of cloud computing. The appeal comes from utility pricing: you just pay as you go. That provides scalability – you pay as you grow – and there is no limit to how large it could grow. But if you want to shrink back, there is no penalty, either. It's the notion of multi-tenancy: multiple tenants or customers would share the same platform so that you can reduce the cost ultimately. It's also the notion of self-service provisioning: if I'm a business user and I want some application, some functionality, I can self-provision the app that I need right here, right now, without having to ask IT to do anything, without having to ask for a big budget to buy servers, or hire a systems administration. And then it's the notion about sourcing: it's the idea that you don't want to own that infrastructure. You just want someone else to own it off my balance sheet, and you want someone else to manage it, as well. Because, quite frankly, I don't have the time; I don't have the skills; and my time is much better spent on working on the core business, whatever my core business might be.

It might be a big corporation. It might be a government. But if I'm a government or I'm a big corporation, the idea of putting my mission-critical business application and data on a public cloud somewhere, most likely in a different country, is either scary or just not possible for legal reason. So I would like to bring the cloud in-house: I would like to bring it in my data center, yet get all the benefits that I listed before.

To do that we put together a solution made of hardware: we bring the hardware, as much as you need; software, the complete software stack with the infrastructure service that gives you the elastic computing and design; the Platform-as-a-Service that you use to build your apps in [force.com] and the applications like [Customer Relationship Management] running on top of that platform, and you can build your own application.

We also bring the people who will manage the system for you, who will do all the software upgrades, who will train your end users, who will train your developers, and who will do a lot of capacity planning so that we know in advance how much hardware you're going to need down the road.

Think of it as Amazon Web Services + force.com + salesforce.com + IBM Professional Services – all packaged, ready to go into your office behind your firewall. We sell that to either the service providers and we sell that to the IT departments of large organizations whenever these IT departments are managed as profit centers so that essentially they have a chargeback mechanism that's in place.

THORSTEN: Why is that?

ISMAEL: Because we need a way to essentially charge the end user. The end user will be a department of the company, and the IT department of that company will just be a service provider serving the needs of the end user. Our pricing model is per end user, so we just charge you a monthly fee per end user much like you would get charged by salesforce.com, for example. So somehow we need a way to get money from that very end user. That's why the IT department has to be managed as a profit center versus just a cost center. Otherwise, they don't have the budget to enable this kind of self-service provisioning.

The idea is that any Director of Sales throughout that big corporation can self-provision an application for his sales team, should he want to do that, so someone has to pay for that, most likely that's the Director of Sales, right? But he does that through an infrastructure that is ultimately provided by the IT department of that corporation, yet operated by ourselves, right?

And so we sell it to these IT departments or we sell that to the service providers, the telecoms, the ISPs, the managed service providers and the outsourcing firms, as well as in many cases the IT subsidiary of very large corporations.

On the telecom side, we tend to work best with those telecoms that have a large systems integration arm. You don't find them so often in the US, but you find them in Europe and Asia. Deutsche Telekom with T-Systems is a perfect example of that, or BT with BT Business Services, Orange with Orange Business Services, NTT with NTT Data, and Singtel with MCS in Singapore. Those are the kinds of telecoms that are very receptive to this model. And the reason for that is by having this IT system integration arm, they have the ability to sell this kind of products to the end customer, and they have the ability to integrate these products with whatever infrastructure the customer already has.

THORSTEN: Your solution together with the workforce that you provide has the brand of being extremely scalable, extremely reliable, and a very high-performance brand for a wide variety of industries and company sizes. So not only for someone who wants to run a database somewhere, or a startup with a few applications, maybe 55,000 users. This works well for 55 or 100 million users. Who is actually signing your check? At this scale, is it still the CIO or is this a CEO decision?

ISMAEL: Ultimately, it's the tenant. Let's take the example of Mitsubishi Corporation. Mitsubishi Corporation is the largest company in Japan. Their revenue amounts to six percent of Japan's GDP, and they're 400,000 employees, 1,500 subsidiaries. To answer your question about who pays us: it's the subsidiaries, and typically the department within the subsidiary.

Let's take Lawson, which is the second largest retailer in Japan, who has got these convenience stores like 7-Eleven. Lawson is a subsidiary of Mitsubishi, and there might be a Director of Sales at Lawson who wants to have CRM for his 100 sales people. That Director of Sales would connect to the Mitsubishi portal and click on the CRM application, which is part of a catalog that has been approved by Mitsubishi Corporation. And assuming that he's got the right permissions, he would be able to immediately provision the CRM application for his 100 salespeople. It would come from his budget, and so essentially there would be a chargeback mechanism where Mitsubishi Corporation would charge back to this division within Lawson, get money from them, and ultimately pay the bill to us that we present to Mitsubishi Corporation.

THORSTEN: So for the division with Lawson it's very much like an app store.

ISMAEL: It's the exact same model instead of having apps like games that run on your iPhone, it's enterprise apps like CRM, business process management, governance, risk management, and compliance.

The other difference is that instead of us being the sole telecom of that app store – like Apple is the sole telecom of the iTunes app store – here we create multiple local app stores or private app stores; essentially one per cloud. So Mitsubishi has a private cloud. They have their own app store on which they put apps that we provide, apps that they build, and apps that are provided by third parties like Zoho – we are one of the first companies that is allowed to resell Zoho behind the firewall, the Zoho Web Office solution –, and apps that are built by the customers. Mitsubishi is putting all of them on their private app store and their end users can consume them that way.

THORSTEN: So not only does your solution lower the [Total Cost of Ownership] for your clients – because now I have a multi-tenant system with shared hardware, software, and professional resources. You also allow a flexible and shared software and application deployment on top of that, lowering [Total Cost of Ownership] even further.

ISMAEL: That's correct.

THORSTEN: And you have software partners who provide applications with much lower license fees and price tags than the usual commercial players.

ISMAEL: That's right, so it's all about lowering the Total Cost of Ownership. It's about accelerating the time-to-Cloud, so making it much faster to be in the position to deliver these kinds of services. For a service provider, it's a way to monetize, to further monetize the customer relationship they have, and it's a really big deal. Deutsche Telekom has 55 million customers. Singtel has 262 million customers throughout Southeast Asia. This is just massive.

THORSTEN: China mobile has 508 million subscribers.

ISMAEL: Right. And as you said, they do not want to be commoditized and be turning to just passive pipes by the software content providers like Google. The idea is to leverage all of the infrastructure that they have on one end and the very solid customer relationship they have on the other, especially the billing system. It's all about the billing system.

THORSTEN: Billing and Charging …

ISMAEL: Leverage this business and operational infrastructure to increase the revenue you get per customer: that's the idea at the end of the day.

THORSTEN: But isn't it difficult to bring such a solution to a telecom whose previous budgeting and underlying business mechanics are very much [Capital Expenditure] focused, not so much [Operational Expenditure] focused? Networks and data centers were purchased and financed with large capital spending and relatively low operational expenses from the few people necessary to run them – or these were even part of a different [Profit and Loss] center. Now everything suddenly is very low upfront costs, instead becoming operational costs as you go. Isn't that difficult to push this into an IT organization?

ISMAEL: Well, there are two things. On one hand there is the telecom that is really good at managing [Capital Expenditures]. To a certain extent, that's their core business at the end of the day: they've got access to credit, cheap credit, and they're very, very good at doing the planning and the deployment of this massive infrastructure and then monetizing it over a long period of time. On the other hand, there's the IT department of a corporation, which is not as good in any way – and finding it increasingly difficult – to get access to this [Capital Expenditure] budget.

So for the latter, for the corporate IT department, it's a very easy story: It's all about reducing the infrastructure costs. And if you can remove [Capital Expenditure] from the equation – even better. So it's a very easy discussion.

With the telecoms, we get in some cases this pushback, and they tell us, *"No, we want to own the hardware."* In that case it is a very easy discussion: we tell them that's fine as long as you buy the hardware that we prescribe and as long as you commit to us to the right level of [Service Level Agreement] so that ultimately we can deliver the complete service with the right level of [Service Level Agreement] that you demand and that your customers are expecting. And so here it puts them in the situation where they really have to think hard about *"Do I really want to do that and am I capable of doing that. And does it really make business sense?"*

And, oh by the way: if you're so good at [Capital Expenditures], actually feel free to buy the hardware and lease it to us because at the end of the day, we don't buy the hardware ourselves. We lease it from a bank, okay, so if you want to play the role of the bank and be the lessor to us, the lessee, that's fine. You know, go crazy and spend your money. If you have so much of it and you can't find a better way to spend that money, go buy the hardware for us. We'll lease it from you, and we'll provide the end service to you. That's okay.

If you want to participate in the management of the system, we'll train some of your best people. We will have trained them to manage this kind of infrastructure. That's fine with us as well: part of the challenge in scaling our business model is to find these skilled resources anyways, so if we can tap into your talent pool to find these resources, even better. It really doesn't change fundamentally the service that I deliver to you at the end of the day.

I'm not selling you hardware. I'm not selling you software. I'm selling you a cloud in a box, and I'm selling you the quick time to market for cloud computing. That's what I bring to the table. The way we bring the different pieces together there is some flexibility here and there.

THORSTEN: So what is possible with Clouds that wasn't possible before? Because virtualization and on-demand provisioning has been available before …

ISMAEL: Five things are new with Cloud fundamentally, and two are really new to Cloud. What's new is the ability of these seven things together.

#1: Utility pricing, meaning that I pay per end user and there is no smallest amount of end users. You couldn't do that with old-fashioned software. You can't go out and buy SAP R3 or whatever they call it now. You can't buy SAP for one user and just pay a monthly fee for that user.

#2 Elasticity: You cannot buy today an elastic server, a server that could grow instantly as large as you need it to grow or shrink back. It just doesn't exist, so elastic computing is quite new and came with cloud computing.

#3: Multi-tenancy is also a new concept that came with cloud computing. The idea that you can share this infrastructure across multiple tenants is fairly new – actually, maybe not so much: you have multi-tenancy with mainframes where virtualization was invented back in the '60s. But that kind of computing almost disappeared with the main computer and especially the PC servers ...

THORSTEN: ... why should you share if you could have your own, right?

ISMAEL: True back then, but looking at it today, it's very easy: If you have your own PC servers, history and statistics show that the utilization rate per server is about 10 percent. If you go to an IT department pre-virtualization, if you go to a corporate data center pre-VMware, average utilization of back office servers is10 percent. So that's massive waste of the [Capital Expenditures]. You waste 90 percent of the [Capital Expenditures]. It's also massive waste of utility, of electricity cost, which is equal to the cost of hardware. So when you amortize equipment – servers over three years – the monthly amount of that amortization equals the monthly utility bill to power the server and cool it down. Essentially you've wasted 95 percent of your investment in [Capital Expenditures] and [Operational Expenditures], 95 percent just to say *"this is my server and I'm the only one using it."* This makes no sense whatsoever, so sharing will essentially increase the utilization and decrease costs by a factor of 20. That's a big deal. We're not talking about 10 percent savings. We're talking about 95 percent saving, so that's the idea of multi-tenancy.

Another idea of multi-tenancy is that hardware operations are certainly expensive, but the most expensive are the people who need to do all the software upgrade of the applications running on these machines. And here the idea of multi-tenancy is: you do these software upgrades once for one tenant or 10,000 – Salesforce.com has 65,000 tenants today – so that's the idea.

#4: The fourth point is the idea that if you virtualize, you can then standardize or normalize your stack, especially of the infrastructure label. So maybe it doesn't matter anymore which silicon you are using; it doesn't matter anymore which operating system you're using; it doesn't matter anymore which database you're using; it doesn't matter anymore which application server you're using. So if all these things don't matter anymore, just pick one of each and standardize across the board. Because if you do that, then the maintenance of all that stuff is very easy. Essentially whenever you get a new version of the OS, you apply that patch across the board to the 10,000 blades that you have. If you do that, the maintenance cost of the infrastructure go way down, and the best example of that is Google. If Google had as many system administrators as they have for every 1,000 servers they have, they would have to hire tens of thousands of sys admins. They couldn't hire them fast enough. And the reason why they don't have to hire so many sys admins, the reason why their ratio of sys admins to servers is so low compared to the ratio in a corporate data center is because they standardize. They've got a single version of the OS. They've got a single version of silicon. And it's across the board, so that's something that's also new with virtualization, which is a key element of cloud computing.

#5: Then there's the notion of self-service provisioning. Today there is not an organization where a business user can go in a portal and self-provision SAP CRM. It doesn't work that way. But today that very same business user can go on salesforce.com and self-provision salesforce.com CRM for 20 users in five minutes. I can use my corporate credit card.

#6 and #7: The last two things are the idea that I don't hold the infrastructure, so I put it off my balance sheet. That's not new. That's the idea of essentially outsourcing. And there is the idea of having someone else to manage it for me, also the idea of outsourcing.

So there are five that are new with cloud computing. You add outsourcing to that, and altogether you've got a new value proposition, and that's the value proposition of Cloud. Now, in all seven elements there is nothing about technology there. It's not about technology. Cloud is not about a technical thing. It's about the business value proposition. And at the end of the day, it's enabling the business with the tools and the apps that they need to get things done without having to worry about the acquisition or maintenance of these tools.

THORSTEN: A compelling proposition. Telecoms are very concerned about this agility because they have to operate the networks connecting these Clouds. A lot of telecoms, as you know, are mandated by regulatory authorities saying you need to maintain networks with a specific standard XYZ. And obviously, as also you know, regulators are sometimes 10 years behind the actual technical standards and best practices. So you have to do some silly stuff sometimes. A lot of telecoms get very anxious when they hear about Cloud because they immediately think it's about connectivity, huge bandwidth of on-demand and unforeseeable data flows around the network. Do you experience that with your clients? I mean: do you come in with your hardware and your people and then the first thing you say is, *"You know what? You need different data connectivity, this one isn't good enough."* or how do you deal with it?

ISMAEL: *[Laughter]* No, this doesn't happen. The way we deal with that is we put our hardware only in a state-of-the-art data center. And so one of two things happen when we go to a customer: Either they have a state-of-the-art date center with so much bandwidth that they will never use more than a tiny little fraction of that because that data center is used for many other applications, or they don't. And in that case we recommend that we put the server in the data center of a service provider they trust. So for example, for a lot of our Japanese customers the servers are hosted by NTT Communications, which is the IP division of NTT, and the professional services are provided by NTT Data. So that's typically the way it works I think.

THORSTEN: I see. Is there anything where you would say, *"Oh gosh, I wish I would have that. I wish we could use this and this and this."*?

ISMAEL: So, the thing we would need today that would make our life easier from a deployment standpoint is for the large service providers that we work with – or would like to work with – to team up in order to deliver a global network of points of presence where I could put my machines. Let me tell you why I need that. All of my customers are multi-nationals, like Deutsche Bank, for example.

In the banking industry there are regulations that say that the data about your customers must stay in the country where that customer lives. So the idea of having a single data center where you put that cloud infrastructure to serve the needs of all your local subsidiaries just doesn't fly. It's illegal. So I need to get access to state-of-the-art data centers around the world on all six continents where I can put my machines. And my customers don't have these data centers, even the very sophisticated ones like Deutsche Bank. So I have to work with telecoms, with telecoms, that give me access to these data centers.

And the problem is there is not a single telecom that will give me access to state-of-the art data centers around the world. And there won't be. I just don't believe that this market is going to consolidate so much that such telecoms are going to emerge anytime soon. What we'd like is the likes of NTT and Deutsche Telekom and maybe Singtel to partner and give me a single face, a single interface, single pricing, or at least single price list, where I can now go to a customer like Deutsche Bank and say, *"Here you go. I've got 256 points of presence around the world in these countries, and essentially we can go wherever you do business."* That would be nice.

And I understand that such kinds of discussions are happening. NTT got close with Korea Telecom. They opened a joint research center in Silicon Valley a year or maybe two years ago, and there are different things happening in that world. We'd like it to happen and materialize faster.

Mobile – you brought that up in one of our conversations before. Mobile is a pain because the market in terms of platforms is so fragmented. If I want to build a mobile app that's going to give the end user a good end user experience, I need to build one for iPhone and one for BlackBerry and one for Android and one for whatever version of Linux and Symbian or whatever. So it's just a nightmare, that doesn't work.

It used to be the same with Web browsers. Five years ago, whenever you wanted rich Internet app that was sexy, you had to do some heavy customization in your Javascript code to the different web browsers. That has improved. Not perfect, but now there are ways to make your rich Internet applications work fairly decently across web browsers, across operating systems for the desktops or the laptops. But for cell phones it doesn't work. And that is a real issue, especially in the enterprise. For end consumers, the market is so large and so dynamic and the investments are relatively small that you can afford to make a version of your game for iPhone and a version for Android, and nobody cares about games on BlackBerry anyway, so you don't have to support it. Maybe not everybody would like that statement, but you get my point.

For the enterprise, you really have to support BlackBerry. You really have to support iPhone much like you have to support the Mac, even though it's a smaller market, and you really have to support Android. And I forgot about Windows Mobile. So. You have got to develop your app five or six times. I wish the telecoms would all say *"Hey, guys, let's stop that silliness. That's not helping anyone and certainly not us. We believe in diversity, but we don't believe in ultra-fragmentation."* Then it becomes counterproductive, and I think now it's really counterproductive. And as an application developer, it's just a massive investment, so you really have to think twice *"Will I build a mobile version of my app because of that?"*, and you shouldn't have to think that way.

THORSTEN: Why don't companies use that rather as an opportunity to give you a web interface, like a very dumbed-down interface where all the function is provided in the cloud? Or stream the application as a video to whatever screen you have?

ISMAEL: Again: It works for desktops and laptops. Because they have a common form factor, a common input mode – well, they used to have, the Apple iPad might change that. It doesn't work for mobile phones. Take the iPhone or your Android phone. A web application on the iPhone is nowhere near as good as a native application. The User Interface just doesn't feel the same. You need to take that into account, especially now that it's not just an Operating System problem. It's also a user interface problem in a broad sense, which used to be that all you had was a keyboard and maybe a scroll wheel. But now you've got touch things, and is it single touch or is it multi-touch? The way you design a user interface is in fact very, very dependent on the form factor and input modes: can you change the orientation of the user interface? What's the implication of that and will I get a keyboard or not? Will it get the stylus or not? Will I get a single-touch or multi-touch?

So all of the sudden, the user interface becomes very, very, very specific, and that's an issue. That being said, I don't know that this problem is going to get solved anytime soon. I think what's going to happen is that a class of devices are is going to emerge as being very business friendly – and granted, the BlackBerry is one of those. But I think that also means that there is a lot of very cool things that could be done with technology that are not being done because of that market fragmentation.

THORSTEN: Skipping ten or 15 years ahead, what will we probably talk about if we have another interview?

ISMAEL: *[Silence]* You see, I don't think I can see that far ahead because if you had asked me the question 15 years ago ... 15 years ago was in '94, that's the year when I got my first web browser, the NCSA (National Center for Supercomputing Applications) Mosaic, later to be known as Netscape. And if you had asked me the question a month before, before I got that browser, my answer would have been very, very different. As soon as I saw that Web browser, I got it. I understood the potential, so I could have given you a meaningful answer, but it was really a matter of essentially the before and after.

So here my answer would be the same. Should I assume that nothing as groundbreaking as the Web browser will emerge, or should I assume that something as groundbreaking will emerge? And in that later case, I can't answer, because I don't know what that thing is. So I have to – the only way I can answer is just by interpolation, by inference, and what I can tell you about maybe not 15 years, but 5 to 10 years: a lot more Cloud obviously, a lot more private Cloud or semi-private, virtual private Clouds, operated by service providers locally as opposed to one big Google cloud, one big Microsoft Azure cloud, and one big salesforce.com cloud. And that's pretty much it.

Recently I saw an article about Cloud saying there is a world market for five clouds, plagiarizing with IBM's Watson was supposedly saying 50 years ago – actually he didn't say it but became a legend – that there was a world market for five personal computers. And that's [censored]. I'm sorry. It just makes no sense. So I don't believe in the public clouds, the gigantic public clouds taking over a meaningful chunk of IT ever. There will be lots and lots and lots of private and virtual private clouds with all the benefits that we talked about. That's granted. The devices will be completely irrelevant, so today I see that most people – not me and you, but most people – have a lot of data on their laptop or their cell. In the future I have none.

In the future I don't really own devices, 15 years down the road. Interfaces surround me everywhere that allow me to do stuff, to be productive wherever I am. So there will be computers, essentially on public access. They're just terminals. They're just dumb, and everything is on a server somewhere. And maybe I'm still carrying a device. Most likely I'm still carrying a device that we call a phone, the mobile phone today, but it's just one of these very, very many devices that I'm going to get. So at home I'm going to get multiple interfaces, multiple devices in every single room, that are going to be used by everybody to just interact with that big Cloud that we're talking about. That's pretty clear to me.

THORSTEN: Kind of the trend from the mobile phone, starting with the first "mobile" phones – the car-battery and suitcase that you had to carry over the shoulder –to the time when you had Wi-Fi access in cafes. Suddenly connectivity and data access became a utility, free to share – or at least freely shared. You still have your personal devices. You need them now to log on, but now the really important part of the game is you have Wi-Fi access.

ISMAEL: You're right. Wi-Fi is actually another interesting thing: It's very hard to remember the world before Wi-Fi. That stuff happened so fast. I mean, it was, what, 2001, 2002 that this Wi-Fi thing happened? And it happened in the span of a year. It was like unbelievable.

Well, I'm not talking about Wi-Fi access everywhere; I'm talking about Wi-Fi access at home or at work. It just came in overnight, and now you never think about using this stupid RJ45-cable anymore.

It's just there, so that's why your question about seeing ten or 15 years ahead – I just can't because you've got these transformative things that happen without anyone predicting it. Of course, the people working on this kind know or hope, but you and I really don't get plugging through. Nobody's plugging through all these potential innovation. And then it happens and the pace of innovation is accelerating, and our ability to adopt it is amazingly good, especially for these innovations that are well conceived and essentially non-invasive.

Think about the time pre-Google where when you had to look for information you had to drive to the library and pull a book like an encyclopedia five years old that was written 15 years prior – so completely outdated, maybe completely irrelevant by now – and look up that information? Now you go and Google, you type your whatever, your keyboard ...

THORSTEN: ...and you believe what you find. [Laughter]

ISMAEL: Yeah, which is another issue, but conceptually we're ready to believe it, so ...

THORSTEN: Touché.

The reason why I'm asking this question is that a lot of telecoms are coming from a history of network rollout and forklift upgrades. And now they are discussion Fiber-to-the-Home, Fiber-to-the-Curb, Long-Term-Evolution wireless, or WiMax. A lot of these investments have a ten to 15 year business case. With traditional thinking, pricing, and services, it's going to take 15 years to amortize the infrastructure. When ISDN was standardized in the early nineties, the business case was calculated for 20 years.

The innovation cycles and disruptions we are talking about just now are very different. Now telecoms have to provide cloud services, hoping or believing that whatever they do with Cloud is going to enable them to do whatever they need to do in 15 years ahead when they don't even know what the business case might be. It's very difficult for them to wrap their head around that. So while for you and I it's OK to eagerly await the innovations to come and say *"Who knows,"* they say *"My God, I can't make a business case anymore for the next 10 years. Terrible. How can I survive?"*

ISMAEL: I don't really think about it that way. I think people who think that way are reducing the value brought by the telecoms to the cables and to the pipes. And it's very dangerous. The value of that pipe is not the pipe itself. Even during the regular landline era the value was not the pipe. The value was the dial tone. The value was the fact that I as an end customer would trust my telecom to give me that dial tone always, any time, whenever I needed it.

So the core, the essence of your business, is trust. And your most valuable asset is not your Capital Expenditure assets; it is not your cable. Your most valuable asset is your customer relationship and to a large extent customer ownership. And that's why you prize your billing system so much. That's why without that billing system you're dead. Because it's the embodiment of that customer relationship, that customer ownership, and it's the vehicle for business. The cable is not the vehicle for business, the billing system is the vehicle for business, the fact that adding a service to that bill is very easy and removing the service from that bill is very difficult. You've got a captive audience, captive customer relationships. It's customer ownership.

So you really have to think about your business through that angle in terms of the services you deliver to the customer, not in terms of what kind of infrastructure you need for delivering these services. The infrastructure, the cable, the bandwidth, whatever it may be – WiMax, LTE, FTTx – really doesn't matter so much. You first have to define the services. And here you can make some assumptions, which are: the need for bandwidth is always going to increase. Always, because more and more stuff is going to go on the Cloud, and you're going to consume that stuff more and more, and it's going to be more and more interconnected. So the need for bandwidth always increases. There is never a point where it kind of plateaus. You could think that the most bandwidth-intensive thing is video. But then comes AJAX and other changes in the user interface, and you need very much bandwidth. So the need for bandwidth always goes up.

I think what you have to understand is that the bandwidth is only a part of the equation. You really need to move up the food chain so that the service that you deliver to that customer brings a lot more value than just that bandwidth, just that connectivity, so that the impact of the [Capital Expenditure] investment that you make on your overall balance sheet is reduced. If you just view your business as the business of delivering bandwidth, of moving bits around, it's very scary. But I think it's much, much more than that, and so that's the way I would approach it. Think in terms of application, think in terms of services, and the relationship to that customer. You definitely have to become an app store. There is no question in my mind that the telecoms must be the telecoms of the app stores that people use …

THORSTEN: … because of the face of the customer, the customer relationship and ownership.

ISMAEL: Exactly! Because I don't define myself as an iPhone user or a Mac user or a BlackBerry user. Remember what I said before: Devices will surround us. They'll be everywhere. And so I don't want to have to link the apps that I consume to the device that I'm using.

Today the app store is linked to the device. It's the Apple iTunes app store linked to that particular iPhone. What I want is much more similar to the way Wi-Fi is being delivered today, where when I go to an airport and I open my laptop and it scans the Wi-Fi networks and there are a couple of public ones and I click on it. And then it shows me the logo of the roaming partners like the AT&T logo, or the T-Mobile logo. And I identify myself as a T-Mobile customer so I click on the T-Mobile logo and I'm on.

That's the way you have to think about the overall portfolio of services that you offer: people are going to identify themselves as a T-Mobile customer or an AT&T customer and will get their services, their applications. And the device doesn't matter, nor the service provider.

THORSTEN: Thank you so much for your time and insights.

ISMAEL: My pleasure.

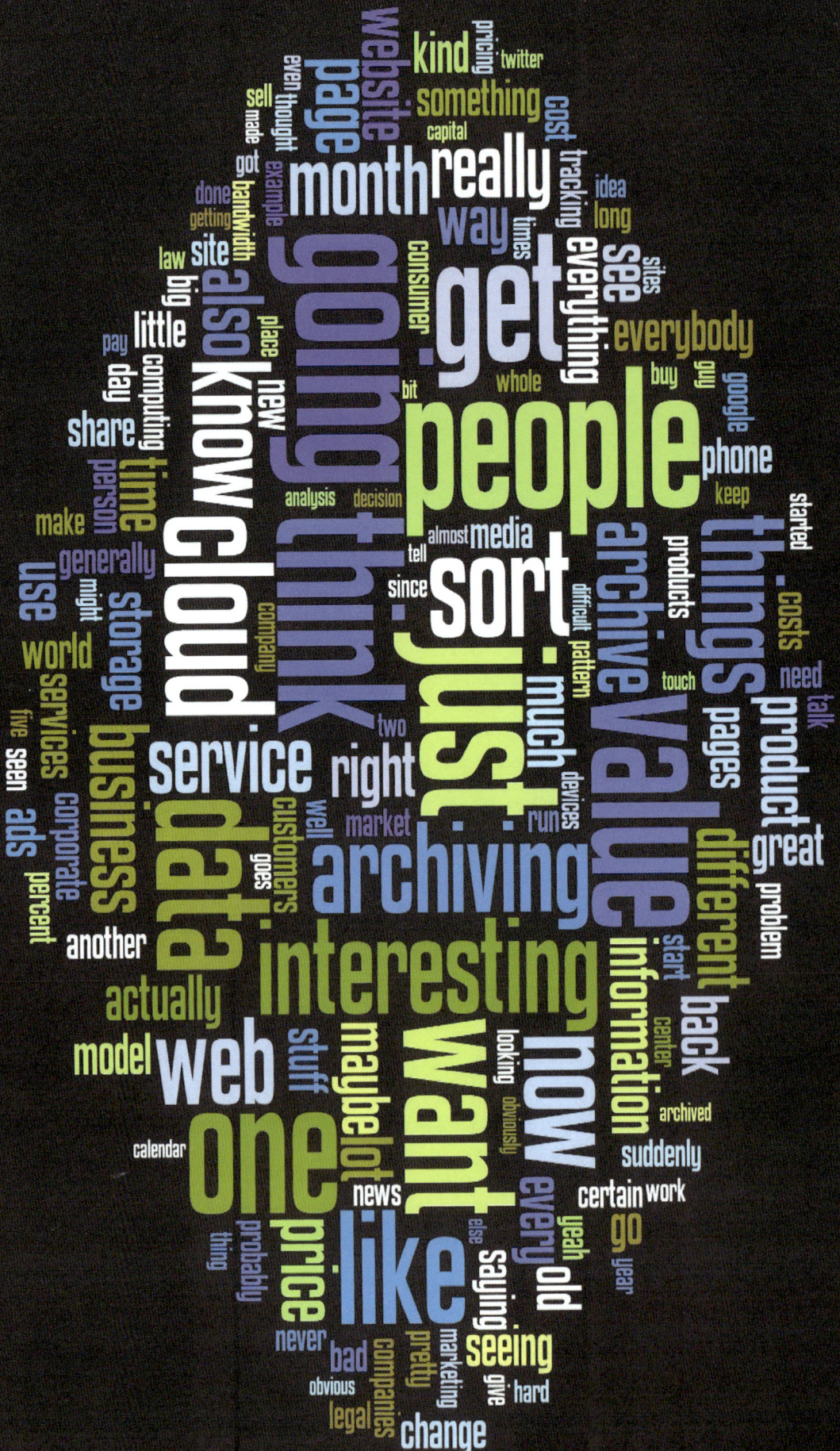

Iterasi
November 17th, 2009

Web 2.0 was the "Web as a Platform", as John Battelle and Tim O'Reilly aptly put it in their 2004 conference. But the transition from traditional computing platforms to the Cloud concept has been unlikely more difficult than the transition from Web 1.0 to Web 2.0. Part of the problem was just exactly that mindset of Web 2.0 and its business models.

Portland, OR, based Iterasi is a web archiving and online media monitoring company that blurred the boundaries between Web 2.0 and Cloud in many ways. In November 2009 Detecon's Thorsten Claus spoke to Iterasi's CEO Pete Grillo in Iterasi's headquarter on the 8th floor of the Pioneer Arts building in downtown Portland about the transition from a typical Web 2.0 startup to a sustainable business with many Cloud aspects.

THORSTEN: Pete, I've been following Iterasi for over two years now. Can you tell us what has changed since you started?

PETE: When we started the company two and a half years ago it was a different world. It was the world of Web 2.0. We were backed by investors and so we were in the middle of what's trendy, and trends change and you have to be aware of that as a CEO. The Web 2.0 model was *"you give it away and then you data mine it for some value"*.

Our line of business is archiving. We're archiving web pages. The initial business models that we were most exciting about were based on *"If I have an idea of what people are archiving – which is sort of like bookmarking on steroids – then there's all kinds of great information I can gather."* And that's aggregated data: These are the articles that are the most interesting; these are the sites that are most interesting; those types contain your ads or someone else's ads. So you just collect a wealth of opportunity there to really say what are people opting-in to, or what are people seeing value in.

Let me explain that. So there's always impressions. There's been impressions since day one: *"What websites are people seeing?"* But if you're asking *"What websites are people archiving?"* then those are a sort of filtering of the most valuable sites people are seeing. So if you thought of every 100 pages you say in a given period of time: if you only archived one, is that good news or bad news? Well it's good news for us because we have the top one percent interesting articles that you've seen. And so there's really interesting information: Who are those authors? What are those properties? And again: what ads are appearing there?

So: brilliant model. I get all excited about it. You reacted positively to it, too. It seems to make tons of sense. But it really didn't happen. I think there's a bunch of factors why Web 2.0 didn't take off. I think one of them is there's just too much of it. There is so much available for free. How do you get on to that user's desktop? How do you get their mind share?

THORSTEN: And how do you get their wallet share.

PETE: Well, wallet share will follow mind share, I think, although I didn't prove it.

So I have the sort of theory that there are only about ten things that people use on their desktop. Those things are pretty high level. Maybe the browser is one of them. And the Microsoft office suite. I have a daytime ticker or calendar app I like. But getting into that short list of things people use all the time is very, very hard. It's very easy to get attention in the beginning, that you can be one of those things as one of the early adopters. We just fell right into that, a lot of initial excitement: *"This is a great service; it ought to be part of every browser, blah, blah, blah."* And then came sort of the drought of *"Okay, how do you stay in people's minds? And then how do you turn it into a business?"*

So we played with that model for a while and then realized it just wasn't happening fast enough. We also played with the ad model. Another wonderful sort of model that will generate a few thousand dollars a month, but it's not anything that you can really build a business on.

I've seen other businesses that actually made a lot of money, that were profitable on ads, and they still weren't satisfied with it. Meaning: it just sort of became an operating business. When will it ever achieve its full potential? It wouldn't. So they're going on to the next level with their business.

THORSTEN: I remember that after the initial enthusiasm suddenly the Cloud came along, and people would ask why archiving, because they could share links over Twitter, Facebook, or bookmark it on Delicious. What is the value in archiving?

PETE: It's so painfully obvious that it's hard to explain in a way – which sounds funny. The web is like the greatest store of information the world has ever seen in one place – or some grand praise of that nature. But it's always changing and things keep disappearing. There's a real uneasiness of things disappearing and so people get that sort of get that implicitly: There's a recipe I like; I can save it. There's my airline ticket receipt; I can save it.

THORSTEN: The old hunter and gatherer story, right? You hunt it down, gather it, and want to keep it …

PETE: You want to keep it and of course bookmarks are only pointers and they fade. But then you get into the nice-to-have versus the got-to-have. The got-to-have is where you get the wallet share. And that – I'm almost embarrassed to say – is taking me longer to find that I would have first thought, but the answers have now become fairly obvious.

Who would really value being able to archive what was said on the web and keep it with a disinterested third party and pay money for that? The obvious markets are legal, government, and third markets I call corporate legal, also the market we were going after in our initial product, which is media tracking.

THORSTEN: For brand tracking?

PETE: Brand tracking, right. Our flagship product is positive press. That market we sort of bunched in – incorrectly, I'll say – PR, media tracking, marketing as being sort of one market. They're really not. PR is what's hot today, what's happening right now. It's been a harder story to sell the value of archiving there, because I really want to know what's happening now.

One of the elements of media tracking is sentiment analysis and engagement. It's like programs that can monitor Twitter. There's like 30 of them on the web where you can find and say that's good, that's neutral, that's bad, and then you can grab all that.

So a sort of logical mind says that you can do that right now with a lot of products. But wouldn't you like to do that for the last 60, 90 or 180 days? Wouldn't you like to see how your sentiment has changed? Gone up or gone down, and then align that with perhaps announcements you've made? You made an announcement and it tanked. But how bad did it tank? How bad did it negatively affect you?

THORSTEN: But why not just keep statistics and data, why time travel back to historic version of web pages?

PETE: You have to have all the data to do sentiment analysis. We archive the complete web page. You could go back and actually post-process things out of the archive …

THORSTEN: Which also brings up a privacy problem: In some of the public archives of your service I saw at least 100 or 200 plain-text email addresses. People captured their Google Mail page, for example, and obviously at the top it states the email address they're logged in with. While it's not your fault, it's an interesting problem to have.

PETE: We thought the public archive was everything and there would be a few people wanting to use it privately. We've switched over our thinking to most people want to save things privately. For a while we were saying we got to save things publicly because then we have data mining we can do on it. But now we give you the choice.

THORSTEN: Back to Brand Tracking: Who is typically buying that service?

PETE: PR companies are usually interested in what is happening right now. We're seeing more interest is in marketing, ad agency sort of companies. Latest news on our website – I think it's up today – is that we're doing business with Wieden + Kennedy. W+K is an ad agency, but they're a media conglomerate. They're seeing value in our services and they talk about why: It saves them time to be able to have this information and deliver it to their customers.

THORSTEN: I like the idea of archiving the complete web page. When you talk to the Department of Defense they say if you want to track a pattern or recognize a pattern, most of the time you only know the pattern after you have seen it. So in order to detect a pattern it's not enough to have a set of five values that you want to track. Because maybe it was the sixth and the seventh and the eighth value that you never knew was important. But later on, after looking at the data for 180 days, they actually established a trend or a pattern. So this is why archiving is very important for actual user experience.

PETE: One of our clients, a sports retail web business, monitors their competitor's website every day. They also archive their own website. The client said it's interesting because he knew how many products he has sold on a given day. But when heading into the holidays, he wants to know what specials he ran last year on his website that contributed to spikes in sales because all he really has are the notes he has taken. He has the numbers and campaigns he ran by looking at his old status reports. But he wanted to see the website. He wanted to see what was presented to his customers on a very specific day. That information isn't generally available.

Some law enforcement agencies want to use this because some bad guy sites you want to be able to actually capture them and have those records. Or on some weirdo extreme groups you log-in a certain way, and then it goes from a family website to this bad guy's porn site or scary terrorist sort of thing. That's kind of interesting.

We've heard from attorneys since we started *"Finally somebody is doing this."* I respond *"That's great; tell me more. What do you do to archive? What do you do if you want to try to find information as an attorney?"* – *"We go to Archive.org and if it's there, great."* – *"How often is it there?"* – *"Almost never."* Nothing against them, they provide a great service but their mission is different. They're recording history on a sort of randomized periodically available copy of sites.

There are all kinds of things that lawyers get involved with. Liability, defamation of character, or just lawsuits where they want to record everything they can. There's the moment of time when they are about to file litigation when we go out and capture everything we can. They wish they could just start archiving their site automatically and have that as evidence.

The other side of legal is the corporate legal side, more of a defensive play: What corporate attorney wouldn't pay X dollars a month or a year to have their website archived by a disinterested third party service? There are frivolous lawsuits that major corporations get two or three or five or ten times a week where your website was supposed to have said this. They want to know for sure when it said that. And there are other ramifications of that inside the corporation: There's the internal Intranet and lawsuits from employees about accusations of different prejudicial behavior. Having a disinterested third party that could archive that information would be a value there.

THORSTEN: Very interesting use cases. A copy of something you own is not a big deal. But if you start copying a competitor's website and share it within my organization – isn't there a copyright issue?

PETE: Copyright laws – my standard speech. Copyright law is two end points and this huge quarry in the middle. And it's open to interpretation by everybody. I've talked to numerous lawyers and debated this on law sites and everything else. If somebody says stop and you stop you're generally okay.

Most people find value in what we're doing. We've had one computer manufacturer say *"You've archived the site, but our price sheet wasn't our price sheet, it was actually a former update sheet. Can you replace it with a new one? We don't want our customers to get the wrong information."* So they were very supportive of what we were doing. Bloggers generally like their information passed around and stored. And we have bloggers who archive their blog with our product as just a separate way to do it. Also, when you archive things publicly they're indexed, they get more coverage so they send their posts and their tweets out through frenzy.

The world is changing so fast, how do you deal with this? If somebody says *"That's my site, what are you doing?!"*, then we just stop. We also have a Digital Millennium Copyright Act (DMCA) standardized form letter we send back saying we're an archiving site, we don't encourage any sort of violations of the law, we'll pull it off. We've probably had 30 or 40 of those since we've been in business. But generally people are very supportive of what we're doing.

THORSTEN: Because if you have website for your own marketing, don't you want your story to go out? Why not having more screens?

PETE: It got a little bit iffy when we were doing ads: we were trying to support our business with ads, we think it's a valuable service. But people are saying you're taking my content and putting ads on it. We realized we were going too far. We don't want to be doing anything like that. We basically shut that off. We experimented with that for a few months at the beginning of the year.

THORSTEN: It's a difficult thing then: Murdoch is trying to pull his content out of Google for the same reasons. They don't think that the value increase through more exposure in Google News or Google Search justifies the value decrease from direct page impression that they get.

PETE: The ying and the yang. I don't know. It will be interesting. I don't want to say this out loud but there could be some changes in Search now that there seems to be a strong competitor and a willingness to change the game.

THORSTEN: Meaning Microsoft?

PETE: Yeah with Bing. I hope so. I hope there's a change in that because let's face it: the consumer always wins in this competition. I hope that the envelope gets pushed a little bit in both directions.

THORSTEN: The interesting story for me is that I can access your service from anywhere in the world over the web interface. And I can issue work orders and monitor the site or give me information about that from anywhere. I don't need any kind of infrastructure. This is what generally people say makes it to a Cloud service, because it's somewhere out there in the clouds. What do you think changed in terms of how this service gets used, how it's applied, or how it's bought, compared to an on-demand model?

PETE: I think it lowers the barrier of entry considerably when customers have nothing to worry about except just being browser based. I think that's probably the biggest single benefit. The other, as you described as "on-demand", is very oppressive. It requires orders of magnitude more complexity. The benefits of Cloud are obvious: I can fix bugs, you don't even have to know – you didn't know there was a bug there, you didn't know there was a fix there, all you know is you didn't hit that bug because I fixed it. New features appear and it's like wow, it does this now. That's great. I'm happy. I get sort of re-enthused, revalidated that I made the right decision, so that's all goodness.

The badness of that is – back to the Web 2.0 – anybody and everybody can create an application. You just need $50.00 and Amazon Web Services to create an application. People just are used to not paying for it. When we had massive installs and you get the new set of DVD's and you have to have an IT guy do the patch every Tuesday or whatever it is: there's this feeling I've gotten more when I deal with sort of Dot Net sort of stuff.

THORSTEN: That's interesting. So the guy that you had to come into you office and had to update your servers was obviously seen as a cost, but now, looking back, it was almost like a value because you saw the guy, a physical person, just working for your money.

PETE: Magic! He puts on the old robes and the pointy hat – the chicken runs around .... So that's where you run into a bunch of problems. With our positivepress product we try to play with what are our entry points. The media monitoring phenomenon is just massive, there's a new products every day. There are free products, you can discount those. Everybody knows if it's free it can't be worth much. It's sort of the consumer mentality. So what's the value you place on this?

We're at a $99 a month entry point which gives you five topics you can track and so on. Well, there are products out there that are $18 a month, there are products that are $29, $39 all the way up to $1500 a month. So I think as a consumer I get really confused: should I be paying $1,000.00 per month because obviously it's got to be good if it's that much? What's a $100.00 a month product compared to a $1,000.00 a month product?

There are sort of different approaches to this in business. When you come into a business a little bit late and you're doing it in software often times they'll provide 80 percent of the value for 20 percent of the price. And you think boy the world is going to just beat a path to your door. That goes up against that perception that if it's 20 percent of the price it can't possibly be something of value. But the other side of that is I think the consumer is confused by this wide spectrum of prices.

THORSTEN: Because you can't immediately put a finger on the benefit that you receive. There is no person with an old robe and pointy hat.

PETE: Yeah.

THORSTEN: And he's going to be here for three hours. And if I would have to work for three hours I know how much that would be worth. Interesting.

PETE: So we're sort of our own worst enemy with these Cloud services, because we place value in this. We're seeing this in our competitors. You place value in this and give a 30 day evaluation. Some people won't even touch it, some people do. You think that the people that don't touch it, they touch it and they get confused, so you try to include support. It's like we'll throw in a half hour of consulting service to help you get started. We see some of our competitors requiring that: *"You can't buy our product unless you sign up for this half hour run through of what the product is".* They don't quite word it like that but that's in essence what they're doing.

So there's one solution, Cloud Computing, which makes everything so much easier, but causes other problems, which is: anybody can get in the game and what's the perceived value.

THORSTEN: Do you think that your pricing model needs to be very flexible even for existing customers? Does the value of your service also change rapidly? And does it also mean that you need to adjust your pricing rapidly? So I'm paying $99 today, maybe next quarter it's only going to be $80. But then there are going to be more premium features. It's very confusing for me.

PETE: As a consumer, do you really want that Chinese menu of choices?

THORSTEN: Probably not, obviously.

PETE: You know we're not hearing a lot of push back on price. I mean we have $99, $199, $399 scales up. We sell capacity because we're an archive so there's storage. Our big costs are storage and captures and care and feeding of course of the database. We're not seeing much push back on prices. That's rarely the issue. I think there's lot of noise in the marketplace on how do I use this thing? And I think that's the function of the decision to go out to a marketplace where it's not a complex sales cycle. Either they get it or they don't. They'll either use it or see the value or won't see its value. As opposed to some of these richer marketplaces where there would be huge value like legal or corporate legal. There's a lot of interesting work going on in the government I'll be seeing on archiving web pages.

THORSTEN: From a corporate perspective, the old idea was to have Capital Expenditures (CAPEX): I bought and owned my servers and infrastructure. But with the Cloud model I don't have that anymore. I have Operational Expenditures (OPEX). Maintenance OPEX in the old model was probably part of another P&L center: Of course you had maintenance, but never saw the bill for it. But now IT organizations buy things differently, and in effect companies need to price things differently, as they in return will affect their own P&L center.

Will the $99 I spend now every month on your product will have the same value in 12 month? Will the value decrease because there are other competitors out there with lower prices, or will the value actually increase, because the longer I use your product the more value it creates? Very difficult for decision makers: how will your costs affect my price and value?

Who are generally your customers? Is it the CIO or the CTO of the media company, is it an employee that signs up ad-hoc? Is there a governing Umbrella over it from corporate management?

PETE: Oh no. Right now there's some IT person or marketing person who's craving this information. We reach that person –

THORSTEN: So even at the media companies it doesn't come like –

PETE: It's not a CEO –

THORSTEN: It's a department decision which says wouldn't that be great so we just do it?

PETE: Yeah. At the price point it's easy to justify. And we've seen that in Wieden + Kennedy where they're using it on one campaign and then all of a sudden they're using it on another and now they're using it in one of their European offices, as I understand. All this stuff is very cool so you want to get into an organization and be viral so everybody is paying $99 or $199 a month, but you're doing it across ten groups, but everybody is happy, and everybody is getting what they want.

At my last start up I was selling the service was calendaring and contacts synchronization, WeSync.com. Palm bought us in 2001. In those days everybody had a handheld strategy, everybody. Enron contacted me and wanted to talk to me about their handheld strategy. Big consumer firms told me that they have 110,000 they're going to give handhelds to. So I'm talking to people but it was all based on the cloud because you're going to synchronize you're going to want to have the data in one place then it's easy to synchronize and shared calendars and shared contact lists. The contact list could be my staff, it could be the babysitters in the community that my wife shares and the calendars, the church calendar, my staff calendar. You know the drill, association calendar. And that's proven to be done now in different ways for different people. Back then it was pretty revolutionary.

I remember talking to a pretty senior IT guy in a major car company. And I'm trying to explain the virtues of having this done and why it's done in the cloud and why we're doing it the way we are. And he's pushing back and then he finally told me, look there's no way I can go tell my boss that your contacts are going to be on the same computer as the contacts from – if it was Chrysler then Ford and GM and everybody else. I just can't do that. The other answer I got a lot of times is: I have $300 million a year IT budget; I have 200 employees including 50 staff programmers maintaining this stuff. It's going to be behind my firewall where I can guarantee the security and the data and all that stuff.

I know that's an old way of thinking but I use the term old loosely here – that old way of thinking is pretty entrenched in corporate America. So I think a lot of what you're seeing is expenses versus capital. Operating versus capital, it's just been done that way for a long, long time, since the dawn of time. There's a certain level of security in it.

So, amazingly – perhaps reverse of what you think – we're actually toying with the idea of an archive appliance that we could sell. Google has Google Search Appliance, now in version 6. I don't have any market numbers nor would they tell me how successful that's been. But the idea is if we could put a box behind the firewall or a set of boxes and a rack and blade servers or something where we could do everything that we do and it was sort of in an old wrap with a ribbon on it – that could be attractive. So we're looking at the feasibility of that and starting to talk to some resellers we're starting to work with and trying to open the government area.

THORSTEN: So what you're saying is that the Cloud becomes more of a hybrid Cloud? A private Cloud? Maybe extendable to a public Cloud?

PETE: Whatever my customer wants really.

THORSTEN: But that's only possible because you are that flexible with your product because it actually is expendable like rubber band

PETE: It's self contained. One interesting example is that we're Cloud Computing to our customers, but we don't use other Cloud services. People have said why don't you use Cloud services for your own service? Well, for one, the price was very high when we looked at it – again this was in '07, '08, and maybe it's come down a whole lot –

THORSTEN: No. It didn't. The price decrease did not go down as much as the volume increased actually. That's my personal experience.

PETE: Well I have the feeling you're right. And things go up in our business about the same because – let me explain: The price of storage goes down, price of bandwidth goes down, and web pages are getting bigger. So our unit of measure has gone up measurably in two years. I don't have the numbers in front of me; my impression is probably 50 percent. Anyways, the average web page has gone up –

THORSTEN: And in about four years 20 percent of all Internet ads are going to be video ads. Just think about that.

PETE: Remember when I web page used to be …

THORSTEN: … four kilobyte …

PETE: … yeah, I was going to say eight, 28, 15 kilobyte. Our average is considered in the hundreds of kilobytes, for an average web page. So that's kind of interesting. But when you talk about the hosting services – the big Cloud vendors that want to sell you the Cloud – they talk about storage and bandwidth. Their pricing models are very interesting, but for delivering lots of little pieces of data astronomically high compared to the price of us doing it ourselves.

I'm also a board member on another little start up that has a very similar but different problem. Going to the cloud for them is a very cost effective way for them to get started. They didn't have to have any IT, it's all there, I can scale it up, and I can buy another virtual server and another virtual load balancer –

THORSTEN: I can actually try things out without the risk of having huge Capital Expenditures.

PETE: Right. The problem is: They couldn't get response times as quickly as their volumes went up. So these Cloud-based data centers are designed for a certain level of service that if you fit in that window maybe okay. I guess I don't even know.

THORSTEN: Isn't that a great opportunity for new players to say I'm going to fill that niche, I'm going to be the high performance super storage company?

PETE: I wonder if their operating costs are maybe more than one would think. So we host in a co-location facility here in town. We pay with everything a little over a thousand dollar a month for a whole rack and all the power. They have a generator and they have four major feeds coming in, and all that stuff. But that building wasn't cheap to build. So I wonder what happens if you have a data center instead of my computers in that building and their supply of basic electricity and bandwidth and one person who goes around occasionally and sweeps the floor , a remote hand that's really just does that.

If you had a whole data center with racks and racks of blades, and all the people it takes to maintain those and view the patches and things like that: I just wonder if their cost to run those things are just more as a percentage as it is for me to run them myself. In other words, it might get more expensive not cheaper.

THORSTEN: That's an interesting comment because everyone is talking about Cloud storage, and people are misquoting Chris Anderson with "unlimited resources" or "zero cost of distribution" instead of "near zero" and "nearly unlimited resources'. There's a big difference.

If I would have the volumes of one of my startups with a mere 500,000 active users every month on Amazon S3, we would spend about $40,000 for storage per month, just for holding the data and transmitting new data. A lot of the times people just push data in, and no one every looks at it again.

PETE: That's what archiving is.

THORSTEN: So we're rather buying network-attached storage (NAS). On the other hand there are new players like CloudScape, for example, providing software to make a Cloud out of your legacy IT. Now does it make sense to do that? Or do you say NAS has a reason to exist?

PETE: I don't know. It all comes down to costs.

THORSTEN: To phrase the question differently: when will be the breakpoint where you would say now it would make sense to have a private Cloud for you? Or to have something that can easily extend upwards and downscale again in case I do need more or less of something?

PETE: It's hard because we have – by sort of my little tiny world that I live in – we have a SAN box that has five terabytes in it. It scales to 18 terabytes. We can also add another one and just wire the bunch together. These boxes are expensive, but not that expensive. I think we got into the first one for around five grand with a few terabytes in it and we've added terabyte by terabyte. It costs us – I'm going to get this wrong – maybe a grand or something like that. So we can scale this thing up in size pretty fast, pretty cheaply. If both those racks are fully stocked I'm churning it at less than 40 grand. And that would probably hold the whole National Archive. When I looked at the hosted services that we're talking about, they were biting me hard on the transaction cost. Not very hard on the storage. I can buy a gigabyte – I think I pay $70. It just doesn't pencil out.

THORSTEN: Maybe their margin is huge. If you look at SMS/MMS for example: sending an SMS costs a telecom something that has six zeros after the comma before it has any kind of number. But you can charge 20 cents. So for Cloud storage companies that for each transaction, for each megabyte they're going to charge you less than 0.01 cents I think it's more like the cost of 0.0000 … – a lot of zeros before any number comes.

PETE: That's interesting because one of the pricing models I saw was 0.1 – maybe it was .015 or .0015 cents per transaction, and a transaction can be a byte to a megabyte. It's great if it's a megabyte but when I'm sending a byte or 4k it really costs a lot then. So you really have to watch. If I was going to move one megabyte blocks all day long this would make sense but they count on you not doing that.

THORSTEN: Salesforce pages lets you expose internal data on external pages. Any change on internal data immediately appears outside, without updating webpage – just press the refresh button. With the machine-to-machine (M2M) business, things are going to change much more rapidly. Inventory is going to change. Data will also be re-purposed for just-in-time customization. Amazon pages are personalized for single individuals for a long time now.

Isn't that an interesting problem for archiving because everyone's page is going to look different and how do you later on argue that a certain element was there when in fact for the person that was viewing the page it wasn't?

PETE: That is what our archive does. It basically says this is exactly the page we delivered to you –

THORSTEN: The archive is going to be my personal view?

PETE: Right. You're personal experience. Very important for law enforcement, where when you log in the right way you get to see the bad guy's stuff. You can't do that with a bookmark.

THORSTEN: Other services just take a generic page view and render it from the view of a user that never visited that site before. But maybe the first screen you get is a big empty one because you never logged on.

PETE: We have certain workarounds but most of it is on premises. If you run on premises we can overcome those problems, but sitting in the clouds we have a similar problem.

THORSTEN: Your service is also social – you can share archived pages. Is this sharing going to explode things?

PETE: I don't know. It hasn't been the case. People are generally – to our disappointment – against this. They're very much interested in the here and now. The river of news is like Twitter: If you don't see my tweet in the first whatever number Twitter decides to show you, you may never see it unless you're searching for it. So not as much value in that as we would have thought.

THORSTEN: When it comes to real-time data and real-time display of that data on the webpage: wouldn't that require constantly archiving the visual page? Shouldn't we co-locating the archiving with hosting companies directly? Instead of you grabbing it over the long pipe, you getting all the data, you having to pay for bandwidth, you could put it directly back next to the hosting rack and instead of having the long line in between, fiber network, …

PETE: That's one of the attractions to the appliance play. I also think certain government agencies don't want this stuff outside. Beyond that, it will be more driven by what the customer wants. I have not thought of it as far as the cost saving for the bandwidth.

Other examples are when it's an Intranet that you want to archive: It isn't going outside of the firewall at all, and you are not going to want to pierce the firewall for the archiving: *"Oh yeah, we have an Intranet, it's all secure stuff, but we're letting you outside to the archiving solution."* that's not going to sell.

THORSTEN: Where do you see the future of Cloud? What do you think is the next big thing that's going to happen? Private Clouds? Pricing? Capabilities?

PETE: The future of computing, I think, is these handheld smart phones. I think everything else is sort of incidental. I still maintain that these will be limited in their computing abilities, and I think your cell phone, your smart phone with the Cloud, is the answer to the future of computing.

THORSTEN: Are you saying that handheld devices are merely displaying and interpreting small chunks of data – the actual work is going to be done somewhere else?

PETE: Absolutely. It has to be for just a whole bunch of reasons: For instance, these computers on my desk: rarely do I lose them, rarely are they stolen. Laptops are stolen. Cell phones are dropped in the water all the time. Having any kind of permanency on a cell phone is ridiculously crazy.

THORSTEN: Some people buy a new phone every nine month anyways. By 2014 we will have about 412 million smart phones world-wide, over 30% will have and open operating system. Only 27% of smartphones are currently priced below $200, by 2014 it's expected to increase to almost 50%.

Of course that will make archiving a little bit more difficult because people will start developing website for specific types of cellular devices – Facebook has a touch, a mobile, a Blackberry, and an iPhone version. So do you have to capture all of these too? And then correlate them? And as a marketing company, do I have to send different messages to different kind of devices, users, and screens? I guess so.

PETE: That will have to be extracted out or the world would just collapse. The burden to deal with that is Facebook's, not anybody else's; Or Google's. You have to have an m-dot application or you're kind of silly if you don't.

THORSTEN: But what I'm saying is if I start having different kind of applications for different mobile devices and I want to do exactly what you said before – what kind of campaign did I run last winter – then you're suddenly looking at a campaign on a smart phone that's not anymore on the market. Form factors might change. Isn't that very difficult suddenly to do data analysis on every changing form factor and user input devices and user behavior? And also the phone you were designing a page for by now had only a quarter of the computing capacity or screen resolution than the phone a year later.

PETE: Or a tenth of it.

THORSTEN: Exactly. Suddenly you might have Flash on the phone; suddenly you might do more data analysis on the screen; suddenly it might be immersive with 3D or something like that. Isn't that like exploding complexity? And what do you think will happen with clients who still have the same need? I want to be smarter and faster to react and go to market, and innovate than ever. So what then?

PETE: Wow, I don't know. I don't know how – I mean you have to sit on the pipe to see what's being delivered down to end users. So monitoring – I'm not saying archiving is the answer to everything. It certainly isn't. Again some people value it more than others. But it becomes the delivery networks that can track that stuff, for ad monitoring, for example.

THORSTEN: Interesting. So you're saying telecoms have a very strong position in that game?

PETE: Well yeah. That's the pipe. I was thinking of the DoubleClicks of the future that are writing the software – it could be the telecoms that monitor that. That would be very interesting.

THORSTEN: Thank you for your time and insights!

PETE: My pleasure.

everything
basic
end
application
trust
stack
move
machines
part
services
regulation
back
big
thing
going
customer
something
stuff
use
number
possible
take
believe
people
aspect
may
example
market
company
model
regulations
today
go
make
partners
networking
become
layer
systems
telecoms
providers
talking
required
bit
understand
access
traditional
lot
get
mobile
way
computing
run
shift
things
Microsoft
pipes
data
done
companies
given
always
infrastructure
find
still
central
provider
every
cloud
network
technologies
government
control
know
step
partner
several
play
space
quickly
hosting
also
really
open
reasons
One
computation
tell
technology
like
service
last
effectively
mind
want
time
give
ecosystem
now
think
need
software
scale
see
large
right
honest
come
virtualization
speak
vs
standard
area
customers
business
different
options
many
new
applications
important
deal
approach
day
running
on-premise
industry
question
mentioned
automation
specific
one
level
ago
carriers
compliance
cost
carrier
years
well
look
much
goes
happen
forth
just
couple
issues
evolve
premise

Microsoft
April 20th, 2010

Microsoft's strategy for the Cloud spans our application, platform, and infrastructure businesses, both for consumer and enterprise. This investment in Cloud services is part of a broader strategy – known across the industry as Software + Services – that brings together the richness of smart, connected devices and the tremendous power of the web.

In April 2010 Detecon's Daniel Kellmereit spoke with Yousef Khalidi, Distinguished Engineer at Microsoft, about how telecoms could enable more choices, flexibility, and opportunities in computing beyond network connectivity.

DANIEL: Thanks for taking the time to talk today. Maybe to start – what is Microsoft's strategy in Cloud computing?

YOUSEF: So, you may have heard us use the term 'Software plus Services'. You can speak of software as traditionally packaged software that we and others have done for many years, that companies deploy themselves on premise, through traditional means. Our strategy is to extend what customers can do and give them an option of: 'Anything they can do on premise by themselves or through integrators, we can offer the same kind of technology as a service.' So customers will have an option of either doing it by themselves or buying it as a service. And they can buy the service from us directly or from 3rd parties. So, our strategy hinges on flexibility of choice. Customers can decide based on business needs, based on geographical location and national boundaries, regulations or whatever the case may be – for a given workload, for a given business technology, deployment, application and so forth. There are options spanning the spectrum from on-premise through to the Cloud, hosted by companies like us or by 3rd parties.

DANIEL: So how do you see that ratio of on premise vs. in the Cloud evolving over the next couple of years?

YOUSEF: To be honest, as you know, this whole area is very new. I cannot give you a specific number, but I can tell you that the dialogue we are having with customers now is…. Well, 2 years ago people were asking: 'What is this Cloud stuff?' and they got answers like 'Cloud is good! Virtualization! Consumption pricing!' and so on. Now they start by saying: 'I want to move to the Cloud!' So clearly, the conversation has shifted. But it is clear to us that not everything will move to the Cloud. We are telling people explicitly: 'We do not believe that everything will move to the Cloud.' Instead, we want to have the option for people to do as they need.

The exact ratio? I do not know. We do not know. Our customers do not know yet. But you can look at the workloads which are currently moving to the Cloud which are more appropriate in the Cloud than others. Around collaboration for example, we have the Microsoft Online Services, BPOS (Business Productivity Online Suite), Exchange, Mail, SharePoint and the like. Collaboration makes perfect sense for the Cloud today. And there are certain traditional workloads that can also move to the Cloud. But even for those, we believe, they will always have connections back to 'on-premise'. We do not believe in all-workload movement. But we do believe in a connection, a bridge, between on-premise and different Clouds out there. Again, I am not answering the specific question on percentage ratios. We do not know yet to be honest, but we will see how the market evolves.

DANIEL: Do you see that there is a lot of room for new players? Or do you mostly see established software and services companies being the dominant force in this market?

YOUSEF: Let me speak to our strategy: Microsoft always believes in partners, and collaborating with partners in creating a big ecosystem for the benefit of everybody. We will not pursue this as a stand-alone thing. We have traditional partners that are fully able to accommodate the Cloud, from system integrators to hosters to other types of partners like ISVs. There are opportunities for everybody in this space. However, the roles change a bit. You will find that hosters for example may only have to move up the stack or specialize. Specialize in a geographical location or in a vertical industry. Integrators will have to invest more and more in Cloud practices that are more meaningful and more targeted at the Cloud. I do believe that customers will want to deal with trusted companies that they know and have strong relationships with. We are not talking about writing software from scratch and throwing away everything, every relationship and every experience here. So the current companies, Microsoft included, who are in the enterprise market, are I believe very well positioned to take the next step up if they are serious about the Cloud and serious about preserving investments and what customers have today. New players will fit in the ecosystem as partners and Microsoft can certainly add additional value. But customers have relationships with companies and they want to preserve their investments as well. So those companies will have to play in the Cloud as well.

DANIEL: Yousef, let's speak a little bit about Cloud applications that are possible today. We have seen this market evolving over the last ten years. First we called it application services provider, then software as a service, later on-demand and now we call it Cloud. What is possible today with Cloud that wasn't possible before? What has changed in the last years?

YOUSEF: A number of things have changed to be honest. Some of the technology aspects which were introduced in the, say, last five years, are key ingredients of the Cloud computing paradigm. And to be specific, virtualization is important. It is not sufficient, but it is necessary. And if you think about it, five to seven years ago virtualization was not really that common. You had it on mainframes and you had it here and there but it wasn't really something that we all, like today, understand. So virtualization was a key ingredient in the picture. Another one is networking technology. By definition a Cloud is something that is delivered over some sort of network, it could be an intranet it could be the Internet. We now have enough bandwidth and required latency around the globe, to enable more and more of that paradigm. But there is another aspect to it that has less to do with technology and more to do with an acceptance. Even a year or two ago, many customers were not able to internalize that they can actually shift some of their workload to the Cloud. So there is an increasing level of trust now that 'Yes, there are workloads that can be moved to the Cloud.' Perhaps, a few years ago, the trust wasn't there yet. By the way, this is not to say that everything will move to the Cloud. Both for technical reasons and compliance and regulation reasons some things will not move there for a long time.

DANIEL: What I always wonder when I hear that is: How important is scale and being a trusted brand to grow in this market? Do you think this is a market that will be driven by a couple of large companies that have a lot of financial power and a brand name or is this a more heterogeneous market, is it going to evolve like the software market? How will the Cloud market be structured in a couple of year?

YOUSEF: This is an excellent question. I will tell you my opinion, but, to be honest, nobody knows exactly what will happen. One can make economical arguments. One can build models to argue that this model will evolve to something like the paid search market with around 3 global players. One can argue for example, that there will be a handful of telecoms that perhaps will become the dominant hardware/software providers, the actual predominant Cloud providers, if you will. That argument goes along the lines of how well you can take advantage of economies of scale – the cost of managing and acquiring a server goes down or the utilization of a server that goes up based on volumes.

Personally, I am doubtful of that scenario. I do not think that there will only be a handful of providers across the globe for a couple of reasons: One is the fact, that we do have national boundaries. We do have government regulations and I do not believe that governments will be happy with having a few mega-companies controlling computing for the whole globe. I simply don't buy it. Today, we have around 300 big telecoms in the world, or so? You know the numbers better than I do. And these companies, frankly, do not necessarily exist because it makes economical sense, but because there are government regulations, because of that human aspect that says: 'I will not, as a government, let all my computing and all my telecommunications go to one central place.' I do believe there will be consolidation in the space, perhaps on the low-end hosting, which will be consumed by higher-end hosting. But the number of competitors will not be low in my opinion. It will perhaps be several hundred in the telecoms arena and some longer tail for certain locations, verticals and countries, coupled with the fact that I also do not believe that we will walk away from on-premise computation. You will always have an on premise presence that will have to be managed through more traditional means, coupled and integrated with a number of Cloud providers. And that number, as I mentioned, I believe to be in the hundreds and not just a handful. That is a personal opinion that can be argued with, but I do believe that ultimately government regulations and national priorities would preclude the situation of having just a few providers out there.

DANIEL: Let's speak about Cloud and complexity. The comparison that comes into my mind here is "driving cars has become a lot easier, that does not mean it has become any easier to build these cars". So, are we making it easier for companies, for end users to host applications, to build applications? And does that mean that for a company like Microsoft, there are now incredible complexities in the back end that have to be managed? Is back-end complexity increasing and will this be an inhibitor for future growth?

YOUSEF: Well, let me tell you what we did with our platform to answer this question. We believe strongly, for cloud computing to be cost effective, that you have to automate things as much as possible. To your point, we are talking about very large scale systems here. And if those have to be managed by hiring lots of people, it will be a very costly and very error prone approach. So the approach we have taken from day one, both for end customers and for us internally is to automate on a very large scale.

And that is why a software company like us can do this kind of stuff. We understand software. We have invested in automation to precisely make it possible for us to have very large scale systems that can work without the extra complexity and without the extra cost. Others in the industry may argue, that they can use services and humans and …. That is what differentiates hosting in my mind from cloud computing. We are not talking about traditional hosting here where you effectively hire an army of IT professionals who have to go and manage every box and every machine. That does not scale. The way to scale is by having the right software architecture with the right optimization infrastructure to make it really scale. And that is why the approach we have taken from day one is to emphasize automation, as I mentioned both for user workload applications and for us internally in how we manage the infrastructure.

And by the way, know that when you have large scale you will have a lot of advantages that come with large scale that will actually simplify management in a sense. Let me give an example: In a traditional system when you have scale-up, vertical, big mainstream like machines, you have to really take care and pay attention to this mainframe. Because it is your only mainframe and it better not crash! In a more Cloud like environment you have thousands of commodity hardware pieces. The basic software architecture that we have and the application architecture can accommodate failure, as you will find other machines available somewhere else. It is not like one or two 'sacred' mainframes that you have to be extremely careful about. You have to still be careful, but there are many, many machines that you can utilize if needed. So yes, with the extra scale you are adding complexities with which automation can help but there are also simplifying assumptions that can be introduced now, because you do now have other options when bad things happen.

DANIEL: Some of our clients are getting worried about a vendor lock-in, where they would be tied to one single vendor for many, many years. They prefer to have an open approach. What is your take on the open vs. walled-garden battle in cloud computing?

YOUSEF: As I mentioned earlier, the area is still new, so we in the industry still need some time to understand the different layers and architectures, that should become a standard now vs. tomorrow vs. the day after tomorrow. That said, there are obvious entry points in these systems that should be open from day one. Customers should be comfortable giving their data to a Cloud provider and, if the need occurs, taking it out. Data import and export are important features from the beginning. You want to assure customers that data lock-in is not an issue and they have options that include data import/export and data portability. So that is an important aspect.

Beyond that, we are talking about computation in general here. The market has a number of OSS and software stacks out there. Those should be supported by multiple hosting providers, so that you as a customer can have options to go to this hoster vs. that hoster vs. that Cloud provider etc., and to effectively run more or less what you are running on-premise, but more effectively in the Cloud model. So choices are important. As we all know, there are costs involved in moving from one place to the other and I will completely admit to that. But to get your data in and out of systems requires that systems are somewhat standard, and of course standard protocols are required to get to them. We are talking about cloud computing, so all access should happen through standard Internet protocols, http and so forth. We should be able to manage them and they should be open to 3rd party tools and for your business partners as well. I think it is a bit early to speak of lock-in, but given the forces out there, I do not think we are going to end up with a few dominant locked-in Cloud providers. I think we are going to have a wider ecosystem out there. And I can tell you that we have customers today who run on-premise and use us to have options like disaster recovery and multiple locations of software, so it is happening already.

DANIEL: A large amount of our customers are telecoms, so let's elaborate a little bit more on service providers and telecoms. Microsoft has always been working with a large ecosystem of partners. What is the role that telecoms should play in the Cloud ecosystem from your perspective?

YOUSEF: There are a number of layers to your question. At a very basic layer, by definition, cloud computing is about networking. You need robust global networks to handle all of this traffic, both for computation, storage, distribution of content and so forth. By definition, it will be the tier one carriers that will handle this sort of thing. So carriers are very important and we partner with them for our own infrastructure of course and our customers do the same. With more data moving into the Cloud there will be more traffic going back and forth, and carriers playing a role tere and having continuous investment in this space for bandwidth and latency is important. Beyond that, depending on what type of carrier you are talking about, some will evolve and take the step to be a service provider. Following the argument that I used earlier, I do not think that it will be a few of them but a few hundred. And some of those will evolve from being pure network carriers to being service providers and perhaps use software from a company like us to effectively become Cloud providers. Beyond that, we will see what will happen in the market. But I do believe carriers will be very important both for the networking part and the service provider part.

DANIEL: We work with all kinds of different carriers. Some of them are highly developed and have been working with Cloud and virtualization technologies for many years. Others have not even touched cloud computing at all. Imagine a carrier that has not done anything in Cloud so far. What roadmap would you recommend? What are potential first steps that a carrier should take?

YOUSEF: Carriers have to pay attention to this space or it will pass them by. What steps can they take to get started? To be honest, they have to understand the space better. They have to broaden their horizon from speaking purely telecoms language. Do not get me wrong here, but we are talking about the intersection of computing and telecommunication. And they need to partner with the right software companies, I believe. Trust me: I understand the investment one has to make in software very well to make this thing possible. There are not too many companies that can afford the software investments that we are making. So carriers have to understand with which software companies they have to partner to get their solutions out. And they have to partner with companies that have partnerships as part of their business DNA to get the job done. So they need to identify the software companies that are leading in the Cloud space and partner with them.

Specific other steps to take will depend on the given carrier. They truly have not done much in this space and still have a long way to go and they have to run quickly. Investing in virtualization should be a given. That is step zero. If they have not implemented the basic step of implementing virtualization, they will have a problem and they will have to move quickly there.

DANIEL: Let me ask you one question which we hear a lot from our clients as well. Telecoms fear that they will step by step become "dumb pipes", providing pure access and not playing a relevant role on the service layer, having no ability to control this layer anymore. This is similar to what is currently happening in mobile, where carriers thought they could dominate the market based on believed core assets like customer relationship, billing relationship, location etc. But by now, mobile is driven by a different ecosystem that does not necessarily involve telecoms the way they would like to be involved. Do you see a danger of telecoms becoming pure access providers? Will there be a "2nd market" that is driven by a new ecosystem of player?

YOUSEF: This is a potential scenario. Unless telecoms play in the market, by partnering, by becoming the service provider, it will become more of a commodity play for them and they will just end up providing the basic pipes. If you look at the stack for cloud computing, at the very bottom you will find datacenters and the usual infrastructure. Then come the pipes and above the pipes there are elements like computation, storage, content distribution, and you have services on this layer as well, reaching all the way up into verticals and given geographic locations and so forth. Carriers need to make sure to not stay on the commodity play level – they need to move up the stack a bit. How high they should move up the stack should be cautiously considered, especially with regards to investing into software, which is why I mentioned that it is important to partner with software companies that have the expertise and software stack to make cloud computing a possibility. I am not arguing for the extreme where you want to take traditional carriers and have them become a software company because that will be both a very high expense as well as outside of the carriers DNA. So partnering is the right way forward.

DANIEL: If you deliver services to a home via DSL or via fiber connection vs. delivering applications via radio access to mobile devices, how would the Cloud infrastructure be different? Will mobile altogether be a different market, or will the same infrastructure and providers serve both fixed and mobile?

YOUSEF: In any scenario there will be a backend involved. A backend is required where the actual computation and storage will have to run. A big part of cloud computing is happening in the back end. And that stuff is quite different to what you have today in mobile. We are not just serving a simple application with a piece of data and you take it from there. There are a lot of things that you want to take care of, from compliance and regulations to heavy duty number crunching, in the back end. So that is what changes the equation for whoever is providing the back end compared to the mobile or any other non-Cloud space you might be in. Mobile is very important, but the Cloud environment is a much richer environment than the traditional mobile space. I really do not know whether it is going to be the same model or not, I really cannot answer that question. From a technology perspective it is more complicated than just the traditional mobile market. Just think of IT issues, of which there are many, and take those and lift them into the Cloud and deliver them as a service.

DANIEL: What are things that can block the road and slow down the adoption process of Cloud technologies?

YOUSEF: There are a number of things here. Many customers at this point are not sure what cloud computing is. One thing that we, as an industry, need to get clarity on is the issue of narrowing down the topic of cloud computing to effectively deliver resources, in the form of computation, over the network, while at the same time avoiding to label 'hosting' as 'Cloud computing', which goes far beyond that. So arriving at clarity with regards to what we understand when we talk about cloud computing is very important.

But to answer the question more specifically: Trust is a big deal for customers and that can mean several things. Most of the Cloud technologies are immature, although they are maturing as we speak. But there is a notion of trust. Why should I trust you, the provider, and give you my data? Do I trust you with running my business in that area? And which part of my infrastructure am I willing to let you run for me? We have to gain customers trust and customers have to be more comfortable in giving up more control. That is certainly one aspect of it.

Another possible hurdle is customers' internal IT departments. IT departments might resist relinquishing control. This is not due to technical reasons or business reasons but just due to human nature. There will be push and shove between central IT and the business units and we will find out which businesses will be willing to move into the Cloud and which will stay where they are today.

Other aspects that need to be considered are regulation and compliance issues. If you look at how regulations are written today, you will find that they are very fragmented between countries. And we as an industry have to work practically with regulation bodies and governments to evolve the regulations to fit the Cloud model. I'll give you a specific example. There exists a specific regulation with regards to the PCI credit card certification process, that demands that the physical machine has to have a firewall in front of it and the exact language of the regulation has been written with physical machines and wires in mind. But cloud computing exists in a virtual environment. That can serve as an example of how the whole regulation and compliance space has to evolve and meet the demand of cloud computing there.

Lastly, we need to have the proper technologies to connect Cloud and on-premise systems. As I told you, we do not believe in a one size fits all here. Some apps will move into the Cloud and some won't and we need to connect them on the network level and on the identity level, and more sophisticated connection technologies are required. So it is a combination of trust issues, regulation and compliance issues, and several aspects of technology that have to be addressed that are missing in the space today.

DANIEL: You mentioned a great last point to elaborate on – the kind of resistance inside of organizations to adopt Cloud technologies. We talked about efficiency. Efficiency will potentially lead to people losing their jobs. So which are the people that might lose their jobs due to increased efficiency? And what jobs will be required going forward?

YOUSEF: This is the story of technology for the last several hundred years. When new technology comes in, certain rules in society or in technology can potentially shift. Without answering your question directly: Look at what cloud computing gives you. It frees you as a customer from having to deal with the low level stuff, such as provisioning machines, bringing the machines up to speed, booting the OS, patching, running the network, and so forth. These functions are performed by people today and those people effectively have to move up the stack and spend more of their time running the business and running the applications above the infrastructure instead of dealing with the infrastructure. Especially the job roles that exclusively deal with maintaining infrastructure will be substituted by automation. And this will cause some resistance.

Other resistance will come from apparent loss of control. Businesses have to go to their central IT departments and ask, beg or plead for their desired application to be deployed and provision the machines and so forth, and there are power play

issues coming into play there. Cloud computing would free up the resources and give businesses what they want very quickly. That can cut both ways for the CIO, but that power shift will have to happen within the company. I predict that the CEO of the company will be more than happy to see the overall IT cost go down. And we are already feeling some of these developments in parts of our customer base.

DANIEL: It sounds like this technology shift will come from top management down into the organization whereas in other area like social networking the technology shift comes from end-consumers and employees and then gets pushed up into the higher levels of the organization.

YOUSEF: I believe that you are correct. The other shift that will happen will be initiated from the departmental areas of the business vs. the central area of a business, as departmental units will often feel a stronger urge to get something done more quickly. Those guys will initiate change and approach Cloud providers directly in order to avoid having to deal with a central IT department to get a new application in place. So the shift to Cloud will come from top-down pressure, but also from the periphery departments of businesses.

DANIEL: Yousef, thank you very much for the conversation and for taking the time to discuss this topic!

YOUSEF: My pleasure.

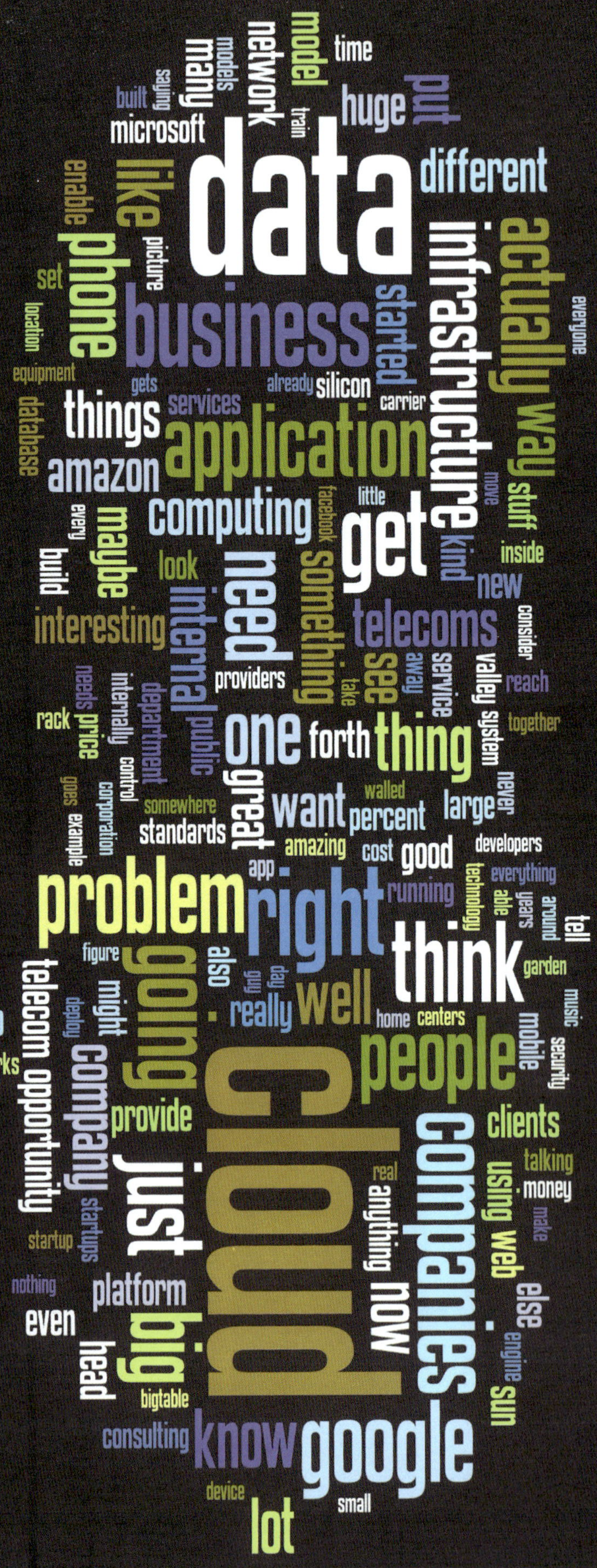
time
model
network
many
models
saying
built
train
huge
put
microsoft
data
different
like
enable
infrastructure
actually
picture
phone
set
business
started
everyone
location
equipment
gets
already
silicon
carrier
way
database
things
services
application
amazon
move
stuff
little
facebook
computing
get
kind
inside
every
maybe
build
look
need
something
new
internal
interesting
telecoms
consider
see
away
service
valley
reach
needs
internally
department
providers
take
system
together
rack
price
public
one
forth
thing
control
great
walled
never
goes
corporation
somewhere
want
percent
large
example
standards
amazing
cost
good
developers
app
technology
everything
problem
right
running
able
years
around
tell
figure
also
guy
day
think
garden
telecom
going
might
deploy
really
well
home
centers
music
go
mobile
security
works
company
cloud
people
opportunity
provide
clients
talking
real
just
companies
using
money
anything
make
startups
startup
web
nothing
platform
now
else
even
big
engine
head
sun
bigtable
consulting
know
google
device
small
lot

Net Service Ventures (NSV)
December 1st, 2009

Radical innovation has been described as a significant change that simultaneously impacts both business models and technology. Located right in the epicenter of the Silicon Valley Venture community, Menlo Park based Net Service Ventures focuses on sponsoring radical innovation. NSV were aware of the "buzz" of Cloud from an early stage, as start-ups and clients began to discuss the then emerging on-demand method of accessing resources. Over time, they have developed a leading understanding of Cloud – how it can be used by their clients, and what the potential could be for various players in the ecosystem.

NSV partner Jamie Allen met with Detecon's Thorsten Claus to share his views of the industry evolution, including some insights into how he believes telecoms can work together to create a new force in the market.

THORSTEN: Can you talk a little bit about the venture investments you make?

JAMIE: Net Service Ventures (NSV) Group is a boutique consulting business, as well as an angel investment firm. Our consulting side drives many of our investments: what we learn in advising our consulting clients, we take and apply it to our investing decisions. We started and incubated about 40 percent of the companies we have invested in. Of those companies, half of them have successfully gone out and raised significant venture.

THORSTEN: Wow. Half of them, that's a great success rate.

JAMIE: We've been very successful in our investing. This success is driven by the fact that we are investing in things that we know very well, things we know very well because they're driven by the consulting we're doing for our strategic consulting clients.

We started out as a consulting business. The investing came along as a secondary thing. We were introducing Fortune 500 clients to the emerging technologies and business models that were happening in Silicon Valley. Through that process we were seeing a lot of companies. And we started thinking, *"Hey, we ought to put a little bit of money in that company."* That's how we got into the investment business.

THORSTEN: When large Fortune 500 companies approached you, what were the common problems they wanted to be solved, with your value proposition being here in the Silicon Valley?

JAMIE: We do what is sometimes called innovation consulting. It's a difficult business. How do you tell somebody how to innovate? Our clients tend to be very large, very successful, and, by necessity, largely focused on maintaining their momentum and protecting their current markets. Their challenge is they don't know what is brewing in the startup community – technology-wise and business model-wise – what is going to come along and attack their existing business. Or, shoot past them, some innovation they may miss completely …

THORSTEN: … creating a complete different value chain composition, break their financial models, …

JAMIE: Yes, that is right. Our role is to understand what their challenges are and their internal strategies, and then bounce it against what we see going on in Silicon Valley, or Silicon Alley, Austin, Seattle, Boston, wherever.

THORSTEN: Portland.

JAMIE: Portland. Wherever. We see the emerging technologies and the emerging business models and say, *"Hey, look. These are things you should consider. You need to be careful about this. You should consider acquiring this kind of technology. Maybe you should develop this kind of a product as an internal venture or working with a Silicon Valley company."*

THORSTEN: So you also consult on organizational structures, recommend carve-outs, etc.?

JAMIE: We have advised that this would be a good thing to set aside and run as an internal startup type of thing. Or we'll do it for you.

For example, one of the companies that we incubated was started with the idea that it would eventually be absorbed into one of our clients. Leveraging Silicon Valley talent, entrepreneurial spirit, and fast to market methods to help a large, slower moving company leapfrog.

THORSTEN: So when you said you want to help your large client to really leapfrog – not only catch-up with things but really leapfrog or have an impact that's radically changing the industry or giving them back their competitive advantage that they have, or keep it, actually – at what point did Cloud enter the discussion with your startups that you start or with your clients?

JAMIE: We were looking at a lot of companies in the video space and in advertising and so forth. And we started to hear from some very early stage companies: *"I'm running this on Amazon web service."* *"I'm running this on Google app engine."* That's where my awareness of cloud computing as something other than just buzz came from. I started thinking about it and looking into it quite a bit. I discovered some people I know were very deep into it and were able to give me some insight. At that point we started advising some of our clients how cloud could affect their business.

THORSTEN: So when you're saying that you saw all the startups suddenly go *"We can create this great product, utilizing something that does not need us to have infrastructure or cost or capital expense and stuff like that."* – How does that relate to a large Fortune 500 company where you have a scale of let's say 255 million customers that are already online? And you're not just 100 people, you're more likely 240,000 people or something.

JAMIE: That's something I spent a lot of time thinking about and talking to various startups and thought leaders in the Valley about. It wasn't immediately obvious to me. Consider the Infrastructure-as-a-Service (IaaS) and Platform as-a-Service (PaaS) models: The obvious economic case for these business models is to provide infrastructure to companies that don't have a lot of cash – for example, startups – it solves their cash flow problem. In the late 90's, early 2000's when I was running startup companies and I needed to build IT infrastructure, I solved that problem by getting companies like Sun Microsystems and HP to finance it for me. To sell me stuff on lease arrangements where I only would pay for what I was actually using and so forth.

THORSTEN: Interesting.

JAMIE: It was creative negotiating. After the dot-com bubble burst and a lot of these vendors were left holding leases on a lot of equipment that was sitting in racks at hosting facilities and the customers had gone away – that vendor financing went away. What cloud computing has done is come up with a different model for financing infrastructure. And it's great. But that's doesn't do anything for big companies: Google, Amazon, a telephone company, or even a big bank. They have cash and the buying power to get the economies of scale, they won't adopt cloud computing for that reason. But what they can do and what they should be doing is using the same approaches in building their own internal IT infrastructure – approaches that are being used by Google, Amazon, Yahoo, and so forth.

When you go to one of these data centers, Google or Yahoo, it doesn't look anything like a data center did in the mid-90's, it's completely different. It is rack after rack after rack after rack of blade servers. They roll in a new rack of blade servers, plug it in, and the software and the configuration gets deployed automatically from some central location. I was visiting a data center in Reston, Virginia, talking to the guy running it, he explained to me, *"When I roll this rack in and plug it in, the guys in Santa Clara will deploy the software and configuration. I don't have any idea about that. I have nothing to do with that."*

THORSTEN: I think we've been to the same data center or headquarters in Reston, Virginia.

JAMIE: It's a completely different model from when I was working for Tandem Computers in the mid-80's through mid-90's. We were selling equipment at banks, stock exchanges and telecoms. The data centers then were completely different – there would be big IBM mainframes over here, Tandem boxes over here, and Sun boxes over there.

THORSTEN: [Digital Equipment Corporation] boxes somewhere else. *[Laughter]*

JAMIE: They were all were running different operating systems and databases. They all had their own applications. And they'd have their own IT staffs and their own application development teams. That's so expensive … forget it! The way that Google built their infrastructure, the way that Inktomi built their infrastructure (acquired by Yahoo), is how the infrastructure underneath the Cloud concept is built. And then you put virtualization on top of that and you get what enables cloud computing. That works for big corporations IT today – they can build their IT infrastructure that way, and get the efficiency and flexibility benefits of cloud – they don't need to buy it from Amazon Web Services.

THORSTEN: But when those large companies try to build these kind of data centers a lot of the times the IT personnel is going to say, *"Why do we need Cloud? We have on-demand already."* And there's a huge amount of automation and maybe the guys from Santa Clara are going to update my service somewhere else. And maybe images are going to be split over several different blades, automatically load balanced. So what's the difference of cloud now? What's the value proposition internally? Why should telecoms care about cloud internally and not just about data center operations? What's the difference between an on-demand business and cloud?

JAMIE: Well, I don't know that there really is any difference. I mean, to a telecom for their internal IT needs, it's all about getting high availability and scalability at a low cost. And how do you do that? You do that by maximizing your buying power with your suppliers. You do that by minimizing the IT staff necessary to run the equipment. You do that by making the development environment very productive. And those are all the attributes that cloud computing infrastructure has. It'll have that, whether it's provided as a public cloud or it's provided internally. And so that's the benefit to a telecom or General Motors or Wells Fargo Bank for adopting these technologies internally.

Most people would say that cloud computing is not a technology, it's a business model. But underneath it, there are technologies that enable the business model. And that's what I think the internal IT, the CIO needs to focus on. The CIO gets rewarded for providing IT capability at the lowest cost possible, and this is a way of doing it. That's why they should get excited about cloud. Now the IT guy down in the trenches might not get excited about it because over time it means a reduction in IT head count.

THORSTEN: But the other interesting view that you mentioned was the operational setup. Like you said: you have small startups, things you try out, like things you first try to fund with seed money. And that, later on, becomes very successful – in your case, over 50 percent were very, very successfully funded later on. Does it mean that Fortune 500 companies should operate in the same way, have small satellites with seed money investments inside the company and see where that goes? And is Cloud the only way to do that?

JAMIE: Well, let me answer that as two different questions. One is: should corporations have entrepreneurial internal startups? I'm a big fan of that. I sort of did that myself inside of Tandem. I took a team offsite, found my own building, and ran it.

It eventually grew into a 200-person team and produced the biggest technology change in Tandem's 20 year history. But it started with four guys in a building by ourselves. And it was very intentional. It was *"get us away from the distractions, from the everyday business. Get us away from the prying eyes of all the executives"* and so forth. And it worked. It worked extremely well. But, key thing: I had the sponsorship of the CEO of the company *[Laughter]*.

And it takes that. Have you ever read [Tracy Kidder's non-fiction book] The Soul of a New Machine? It was written in the mid-80's or something. But it's about the building of a new system at, I think [Data General Corporation]. I knew some of the guys that were featured in the book. It started the same way. It started as a skunk works down in the basement. But it had the sponsorship of somebody high up in the company who could pull strings and protect them. That model works, but it also tends, sometimes, to encounter a lot of resistance. The people that are not part of the startup team are resentful; they can put up barriers, so it needs sponsorship. It can help a big company do things quite quickly that they might not otherwise do, which is a problem telecoms have, inherently have, but all big companies have.

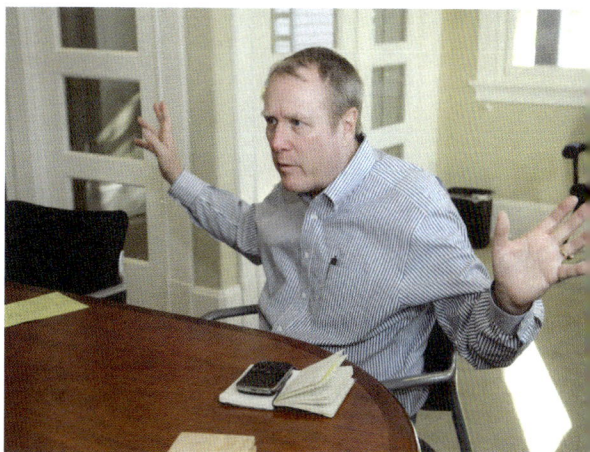

The second part is: does Cloud have anything to do with that? It can. Buying infrastructure from Amazon Web Services, Engine Yard, Google App Engine, whatnot, can enable a small department inside of a big corporation to quietly do a skunk works. The department manager could put it on his credit card and he can expense it on his expense report. His manager might not even notice or know what it is. So it can enable things to get done that would not happen if you had to go through normal channels of acquiring equipment through the IT department – all the sudden you would have all these barriers: Are you using an approved operating system? Are you using an approved compiler? And is this the database you're using?

THORSTEN: Physical security.

JAMIE: Yes, security. Something that you could spin up and try for a few hundred dollars on an Amazon Web Service, if you had to do it through the IT department of a big corporation, would cost many tens of thousands of dollars.

THORSTEN: Isn't that silly, though?

JAMIE: It's very silly, but on the other hand, it's not: the IT department is tasked with making sure that anything that gets deployed can be supported for the next 5 to 10 years, and the cost will be contained, and so forth. Their job is not to enable entrepreneurs. Rightfully so. I think the corporations should allow people to experiment … some Silicon Valley companies have this concept that 10 percent or

20 percent of your time should be spent doing something that you think is neat – Google is the flagship of this. Using cloud computing is a way to enable these kinds of entrepreneurial things to happen inside of big companies.

THORSTEN: Wouldn't it be more effective – especially for telecoms that already have data centers, with sensitive data about their users, with sensitive products – to have an internal cloud facility to keep exactly those tasks inside? Currently people basically put things into Salesforce or rent themselves some Sun space, you know, and suddenly customer data might leak into someone else's system – whether that's allowed for the regulatory authority or not. Now even though the Sun cloud and the Microsoft cloud might be extremely secure, a regulator is not interested in how secure it is: he says it's not allowed, you can't leave your premise. But in this case, it does. Wouldn't it be much smarter for a telecom to provide that internally, those kind of things? And then it's also more cost effective.

JAMIE: The issue of privacy around data and so forth is a very serious issue. That is one of the reasons I do not see big corporations in general, and telecoms in particular, using public clouds for their core business. I would certainly not advocate the small department entrepreneurial team using a public cloud take their private data and put it out on the public cloud. But it does happen. People put their corporation's customer data on their laptop, and the laptop is stolen at the airport. It happens all the time. Social security numbers, people's medical records, and so forth. I personally have received letters from companies I do business with saying your data was stolen. And so that's a real problem. And I agree with you that this private customer data should never leak, should never be allowed to leak.

But there are ways to deal with private data in a test environment. You can anonymize the data before you use it in a test database. The people who know what they're doing, that's what they do.

Providing an internal cloud that allows people to have control – the ability to provision computing resources and deploy an application on their own without involving IT – that's the ideal picture. But, I have to tell you, that just last week I had lunch with a guy who's a very senior engineer at a very big IT providing company who had an internal cloud. It had all the attributes of a public cloud infrastructure except that it didn't allow the granularity control – you had to go through the IT department to get your application deployed on this cloud, it was crippled in that way. I think that the "control thing" is the hardest thing to overcome, that is where IT says, *"Oh, wait a minute. That's my job."*

THORSTEN: But isn't it funny that a Salesforce is able to provide external developers complete and satisfactory development environments. At Salesforce I can roll out applications whenever I want. The only thing to have is a third-party certification what the app is actually doing, a function that Salesforce actually provides through an independent third party certification program. But then you can actually do that. And you have all those big resources, resource pool, and you share that in a fair and sound and safe way. And now the telecom's internal IT says *"We're not going to do that. We don't trust that."* Isn't it interesting that they don't do that?

JAMIE: Well, it's culture. It has nothing to do with technology or good business practices. It's pure, simple culture. I can tell you many stories. You know, your parent company – I'm sure you have many stories as well.

I think the real opportunity is telecom's adoption of cloud internally. And I think that's critical for all of their internal IT infrastructure in order to contain costs and be more flexible. In a nutshell, I think it's absolutely critical that any large corporation have a strategy for evolving to a Cloud-like internal infrastructure and operation.

But I think the huge opportunity for the telecoms is to become public cloud providers, to compete head to head with Google App Engine and with Amazon Web Services and with Microsoft Azure. The telecoms know better than everybody in the world how to operate large infrastructure 24 hours a day, 7 days a week. And how to negotiate the best prices from their vendors. Telecoms are experts at managing the vendors and getting the lowest price, and operating infrastructure that's available 24 hours a day, 7 days a week. So why not? Why shouldn't they become Cloud providers?

One of the biggest barriers to broad adoption of cloud computing is the lack of standards that enable portability and interoperability. How will those standards get set? Maybe Amazon and Google will sit down and say *"let's set standards"*. I doubt it, they are going to compete head to head – Amazon and Google and Sun and Microsoft – I doubt very seriously that they are going to agree to set standards. But what if AT&T and Verizon and BT and France Telecom and Deutsche Telekom and Telecom Italia talk together? Four or five or six of these people get together and say, *"You can own cloud computing in your region, and I'll own it in my region, we won't compete with each other head to head within our regions. Let's set standards that enable portability and interoperability across our clouds so that the application developer, when he says, 'Where am I going to deploy my application?' He'll choose to deploy in a telecom provided cloud because he knows when he deploys it in the U.S. he can deploy the same thing in Germany or in France or wherever.*

THORSTEN: Now, that's an interesting dilemma, because Google says you can do that exact thing right now already – as long as you stay on the Google cloud. And Sun says you can do that exact thing right now – as long as you stay on the Sun cloud. And the problem is that all these companies, Sun, Microsoft, Google, they're all global companies, while most telecoms are actually quite local. They're mostly confined to their network footprints. A Deutsche Telekom has their core network assets in Europe. Yes, 57% of their revenue comes from international business, with – I guess – probably about 23 billion US-Dollar business in the U.S. alone. But they're not that global like a Google, who is ubiquitously accessible with the same or similar business model around the globe.

Telecoms can't see that as an asset. So you just said running the data centers is a big asset, and having processes in place, and having control in place, and doing vendor negotiation in place is a big asset. Telecoms see that as a disadvantage. They say they have inflexible and complex processes. They have all these vendors we need to manage. It's interesting, right, because there's this dilemma: how do you change this kind of mindset or the culture? It's actually a real opportunity and not just something that's like a heavy weight on your leg that you have to take along.

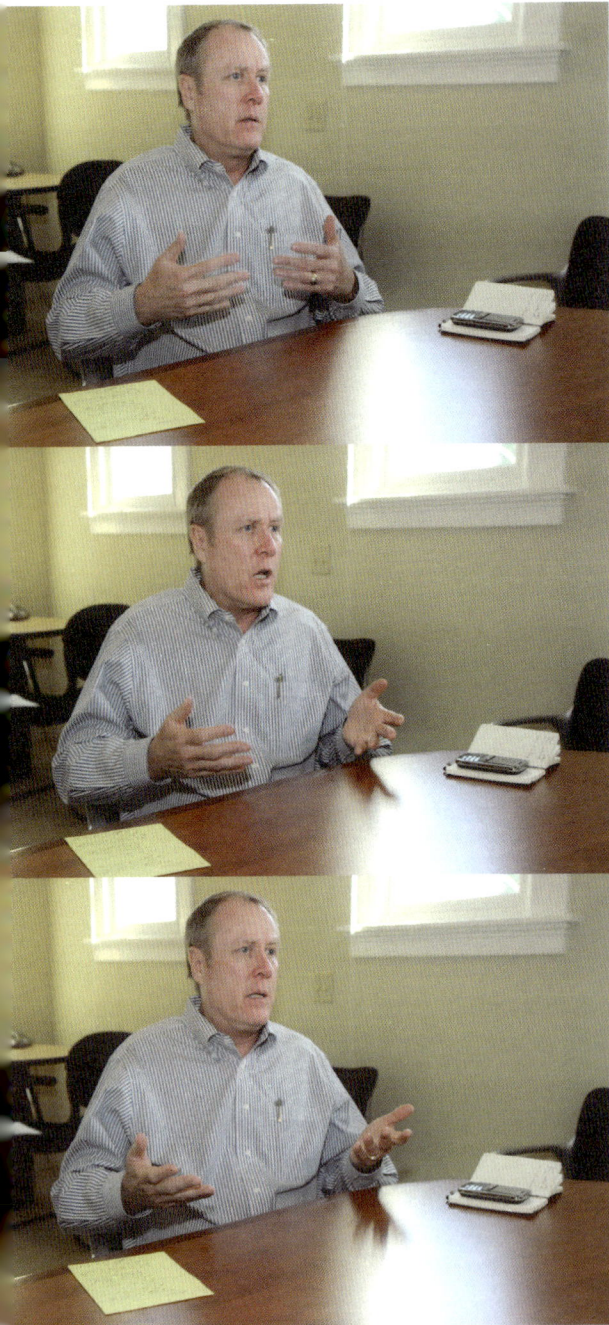

JAMIE: Right, you put the finger on it; that is the dilemma. Their very strengths, their expertise that is their core strength in many ways is strangling them. What it took to build the phone company in the first half of the last century, the engineering that it took, is not appropriate today. But in many ways that's how they still think: *"we have to spec everything. We have to guarantee that it's built this way and it's labeled this way."* And on and on and on and on and on and on. No. You need to think about how you could build a reliable system out of unreliable hardware. Just put a lot of it together. You put a lot of cheap computers together and you get a lot of computing power and it's very reliable. You need to think differently. If they could think differently, and bring these great skills and expertise together, then they could do amazing things. In the technology world, what are the standards that have changed the world the most in the last 50 years? TCP/IP, and DSL. Right?

THORSTEN: Maybe HTTP, in terms of the common infrastructure for websites.

JAMIE: Absolutely, also on the list. But the point is: the telecoms are responsible for one-third of the list, DSL, with a strong influence and participation in the other two standards. Well, why can't they do the same for cloud computing?

THORSTEN: Interesting point.

JAMIE: I think there are huge opportunities here for them because as a cloud computing provider they could provide the public cloud and they can compete head to head against Google and Amazon Web Services. But they also could provide a cloud inside the Walled Garden. I have to note, I'm the guy saying tear down the Walled Garden, I hate the Walled Garden. However, there is an opportunity for the telecoms to provide a cloud based an application platform for their walled garden – good role model would be [NTT DoCoMo's i-mode] – an application platform that would allow application developers to deliver services to the telecom's broadband and mobile customers. A platform with integrated billing, customer support, and high security for people that are security conscious. The telecom can say, *"Hey, use this service. The data never leaves our network. It never goes out on the Internet,"* kind of thing. There is a market for that, and nobody else in the world can do it.

You offer the application developer the opportunity to leverage the telecoms brand and to reach those 20, 30, 40, 50 million subscribers that the telecom has. Where else can an application developer go to reach populations like that? Google is not going to help them reach 30, 40 million – Facebook does. Facebook helps you reach 300 million people. Access to the user base is the reason why application developers build on Facebook, the whole reason why. It's because you can reach those millions, not because it's a great platform. It's actually not a great platform, it's just a so-and-so platform.

THORSTEN: But it doesn't need to be a great platform if it serves your needs.

JAMIE: That's right. Well, the need it serves is the ability to reach a huge population of users. Telecoms can provide that capability to application developers.

THORSTEN: But then obviously they're afraid that the network is going to collapse, etc. … you know the whole story … the chatter on the mobile network, research showing that only 30 percent of the bandwidth create 80 percent of the chatter. Because constantly apps keep querying, *"Do you have new mail? How's the weather? Do you have new mail? How's the weather? Do you have new mail? How's the weather?"* And all this kind of stuff. So they're very scared of it.

JAMIE: It's hard.

THORSTEN: The big white elephant is here. Everyone is talking about it. No one knows what's happening …

So let's talk quickly about stuff that you do see. You just mentioned that some of the ventures that you've started are looking into video and advertisement. Can you tell us anything about it? Or is that all stealth mode?

JAMIE: Well, most of it you can see on our website. So you can check them out. And they're very, very public. Go check out the [ShortForm TV application] on Facebook.

THORSTEN: So you said a lot of the companies use Cloud stuff. Facebook is basically a Cloud platform, because your application runs somewhere and has this wrapper of social networking around it. Is there something you would say, *"Gosh, I wish I would have XYZ right now on top of that."*?

JAMIE: There are actually all kinds of things. In most of the clouds, databases are a problem. If you're going to Amazon Web Services – besides being too slow and too expensive to scale – you can have a MySQL database. And that's great. But when the instant goes away, the data goes away. That's a big problem. You have to stand on your head to migrate your data over to S3 or someplace else. It's a serious problem for a serious application. At Google App Engine, you can't have a SQL database. All you have is BigTable. There is nothing wrong with BigTable, it is a fantastic database for its purpose. But for one reason or another, if you need SQL, you can't get it.

THORSTEN: But what about specialized services like Terracotta, for example, or Greenplum, who have specifically targeted database. Oracle is now entering the cloud with the [Sun Microsystems] deal. Is that not an option? Is it too expensive? Is it too difficult?

JAMIE: They are options but you run into that problem of where, you go here and you get this set of features, and you go there and you get this other set of features.

THORSTEN: So that's a standardization problem again.

JAMIE: Yes. If you build something here you can't build anywhere else. I have an application where BigTable is the right database. It currently is using SQL. It should be on BigTable. And so I wish BigTable was available when the application was originally developed. But we're now thinking, *"Well, can we move it over there?"* The problem is it's a Ruby on Rails application. Move it over to Google App Engine and it's fight time. There is a Ruby on Rails implementation but it's not good. The performance is terrible. So these are the kinds of problems that really limit what you can do in the near term. And then Microsoft Azure: talking to people at Microsoft it sounds really good what, they're saying they're going to do. But, again, it doesn't support a Rails environment.

THORSTEN: But Microsoft Azure is interesting because you have so many .NET applications running on legacy hardware. And obviously porting the same complexity just into the Cloud will not solve you any problem. It will actually create more problems because you have the same complexity that can scale faster.

Great. So it gets more complex even faster than before. Bad idea. On the other hand it is actually possible to migrate. You're not stuck. You can actually do that. And maybe from there, after you have done that, now you say, okay, let's take it apart and step by step we're going to do it better. But you actually can move it to somewhere else. Very attractive.

JAMIE: Microsoft's got a built-in market. They provide a Cloud that supports .NET apps with SQL server and with all their development tools. Why would they do anything different? If I was advising them, that's what I would tell them.

THORSTEN: You have the existing channels and developers and everything. Nothing changes on that.

JAMIE: So that's absolutely the right thing from them to do. And they'll do it. The great thing about Microsoft is they'll stick with it until they get it right, and it works really, really well. It might take them five years. But they'll get it right. They have amazing talent.

THORSTEN: Is there anything particularly spectacular that you recently did where you say, *"Wow, this was the best ever for telecom"*?

JAMIE: Well, we did a thing called My Digital Life. You can actually see a little bit of it on our website. The mobile phone is this amazing device. It's with you 24 hours a day, 7 days a week. It's amazing how much data is on them. Not just data that the user generates, the user's phone book and the photos he takes. But all this silently collected metadata: location information, a record of every call you make and receive, every e-mail you send and receive, and so forth. We built an application that collected all that data and brought it into the cloud, and it organized it. We find the threads of conversations that went on for weeks or even months through all this data. It allows you to explore these threads of conversation -- some people call that life-log. It was an extremely neat thing and very powerful. But we couldn't figure out how to commercialize it, so right now it's in the ice box.

THORSTEN: Very interesting. We had a lot of different business models and cases, something where location information of a telecom could be used to evaluate how valuable your commercial real estate is: You know that each day 400 phones move by that space. And you know how fast they move. So you know whether people are walking or driving. Does the commercial signage need to be big or does it need to be small? Is the traffic mainly facing into one direction? Should you actually advertise offers? Should you cater to walk-ins? Would you rather cater to the web? You don't need Google or Navteq or anything like that to know whether this is a one-way street – everyone is just driving in one direction, right?

JAMIE: Figure it out.

THORSTEN: And if you have this kind of information, you give it to a real estate agent: *"Look, I can prove you how many people you're going to see. Every day. This is proof."* Right? And immediately he can show his clients. *"This is not just what I'm saying. I can get you data from the last five years if you need it."* It's a very, very interesting business model around location. And that's where, actually, I think Cloud is very interesting. Because it's not just about my thread or my data – you can anonymize all data and leverage all location.

JAMIE: That's right. There was a company that was trying to do that, but it was dependent upon installing clients on the phone.

THORSTEN: Yeah.

JAMIE: And that's not going to work. What works is getting the data straight from the mobile network. Because then you can get all the data. And then you can really see: Oh, the traffic on interstate 280 has stopped right here. It must be a problem there. You can tell that faster than any other way. Even faster than people calling 9-1-1.

THORSTEN: Right. Suddenly everything stops. Or suddenly everyone is taking pictures.

JAMIE: There is amazing intelligence that can be captured from all this noise if you could just get all the noise in one place so you can analyze it. I think that's a huge opportunity. But there are privacy issues that need to be taken into consideration.

THORSTEN: But wouldn't a telecom be the best guy to have the brand and recognition to be secure. I'm talking again about the Walled Garden. Nothing leaves my network, right?

JAMIE: No personal data goes anywhere. It's all anonymized and so forth. The other huge opportunity I don't think anybody is really dealing with effectively today is the distribution of personal data. The amount of data that is on this mobile phone – just consider your calendar. I want to see my calendar on my mobile and my PC. Maybe at home I don't have a PC, and I want to look at it on my TV. Today, I just can't do it.

Consider music. I want to have my music library at home in Apple Lossless Format or the closest version of mp3 -- huge files that I could never put on my iPod or on my phone. I want to have a subset of my music library on my phone in an appropriate format, and I want to be able to rotate it on and off. There's no good system for doing that, for managing this distribution of personal data. I think that is a huge problem, and where there is a huge problem, eventually somebody will figure out how to make money solving it.

THORSTEN: Not only is it important to know how to get the data on there but if you keep drawing the storage curve of the iPod, for example, in 2012 you're at a terabyte. It doesn't make sense to manage a terabyte storage on a device yourself. Maybe if we have 3D data and HD and you could connect it to the TV and play stuff off it, maybe the terabyte fills up more quickly. But aren't we all the hunters and gatherers. So everyone is going to have one terabyte of data on there because you might need it one day. I mean, look at my phone. It has 16 gigabytes of data, full with music. How many songs do you listen to? Maybe 40, 50. [Laughter]

JAMIE: You should talk to my daughter, who is an adult. Every couple of years she buys a new iPod because she needs more storage. She keeps her entire music library on an iPod, plus she keeps a lot of video on there. She consumes whatever storage is available.

THORSTEN: Other startup companies I have seen have transcoding facilities, or set-top box providers trying to do transcoding and intelligence gathering in the home; uploading stuff from the telecom to the homes, basically extending the Cloud onto the subscriber's premise. Then my home set-top box knows that I have a T-Mobile G1 phone and you have a Blackberry. So it could transcode the data exactly for my device or your device. That would be nice, too. But the [Digital Rights Management], how do you solve that problem? I'm also not always online, right? If I'm up on Sonoma, I lose signal all the time. I sometimes don't even have reliable 3G connection in the middle of San Francisco. I can't stream everything.

JAMIE: No you can't. Putting everything in the Cloud is, is, …

THORSTEN: The wet dream of Google, maybe. Though they did invent [Google Gears].

JAMIE: It's just not going to work, nobody is connected all the time. If you commute on a train in New York City or Paris or London or Tokyo, you're underground a lot. Tokyo is as good as it gets and the coverage is mostly only in the train stations, in between the train stations you're disconnected. We learned this during a trial of My Digital Life there – people said they wanted to write blogs, so we provided them the ability to write blog entries, journal entries, with a web service on their phones. They said that is not good enough because I want to write it when I'm riding the train, but I'm disconnected. It means you have to put an application on the phone that can cache the data for access when you are disconnected. The way I think about it is personal data should all be stored in the Cloud, and every other device should be just a cache, but a very intelligent cache. I shouldn't have to tell the system what pictures I want to have on my phone. It should figure it out by looking at what pictures I look at.

THORSTEN: So all this kind of intelligence about what do I like, what do I not like, when do I look at what stuff – it should not be a manual process.

JAMIE: Oh, no. 90 percent of it should be automatic. However, I should be able to explicitly configure it to never remove the picture of my grandson from my phone. But generally speaking it should be automatic. For example, if it's a picture I recently took it should probably stay on the phone for awhile because there's a probability I want to show it to people. If it's a picture that I have looked at recently it should stay on the phone, there's a probability that I'm probably going to want to show it to somebody else or look at it again. But if it's a picture that is three months old and I haven't looked at it, and it's already been uploaded to the Cloud – get rid of it. And then, if I do download it from the cloud again, don't send the four megabyte file down to my phone, I only have a little teeny screen. Transcode it and send a little teeny picture down.

I think this problem of managing peoples data distributed amongst many devices is a huge problem and, therefore, is a huge opportunity. The problem is how to make money out of it.

THORSTEN: The underlying problem is also finding a commercial model where Cloud providers or software providers or telecom providers are actually able to pay the amount a telecom needs to create the network they need. You just mentioned your users that want connectivity in the train. Who's going to pay for the network build-out and what's the commercial model? What are the mechanics you need to enable a telecom to do that? And it obviously has to do either with price reduction of operations or with wallet share, with increasing revenue from somewhere. So I think it's a big problem right now. No one has figured that one out. And that's an underlying enabler, right?

JAMIE: That's true. That's absolutely correct.

THORSTEN: Of course, in Germany we also had the big auction of UMTS spectrum for back then 71b US-Dollars. And you haven't even built a single tower or backhaul, yet. How do you get that money back? It's very difficult.

Even if you put fiber in the ground, every home passed is a 1,800 US Dollar bill or something like that. And then the regulator might say, *"Great, now we have fiber. Let's open that for everyone else, too, to a fair price. And I say a fair price is 20 Euros per month."* So you need to have 60 months of service just to cover your network deployment costs.

JAMIE: The opportunity in Europe – this amazing thing happened in the U.S. about 1994, 1995. McCaw Celluar developed these independent businesses around the country. Northern California was actually owned differently than Southern California and New York. They were all different. They ran different technologies and so forth. And McCaw owned like 50 percent of each one of them. And I'm trying to remember if it was right after, I guess it was right after the old AT&T long distance business acquired McCaw, they came out with the one nation price. You could call anybody in the nation for one low price. More importantly, there was no roaming. You could go anywhere in the country with no roaming. That's the opportunity in Europe. Some upstart is going to come along and figure out how to provide roaming free coverage all over Western Europe. That's going to be very disruptive.

THORSTEN: Google could sell a mobile phone that does exactly that app-based, connecting seamlessly over cellular or Wi-Fi networks, like my wife's Blackberry. And you price it like the [Amazon Kindle], with an all-inclusive one-time fee for the lifetime of the phone, plus up-sell hardware warranties. Then they could collect personalized and localized user information Europe-wide, to better control the value chain from ad Dollars to ad display …

JAMIE: That's a good comment.

THORSTEN: Thank you. Well, thank you so much for sharing your thoughts and insights, great talking to you, as always!

JAMIE: My pleasure.

carriers

Cloud

content

companies

network

NewBay

services

see

OEMs

opportunity

APIs

consumer

many

major

carrier

like

provide

use

data

years

solutions

around

parties

Content

Networking

cloud

user

Service

User

example

think

need

devices

platform

make

develop

also

going

Social

Facebook

paradigm

Twitter

metadata

huge

access

Obviously

technology

consumers

using

play

even

across

LifeCache

Laughter

globally

already

better

just

next

offer

control

customers

networking

compete

providing

obviously

get

long

well

interesting

quite

change

segments

photos

though

rather

right

sure

differentiate

activity

biggest

bit

downtime

Digital

Cloud-based

conditions

strategic

Ecosystem

Gateway

flexibility

areas

resource

five-nines

now

allows

applications

introduced

standard

manage

Carriers

space

standards

role

days

want

really

today

screens

Today

enables

terms

share

trusted

customer

Generated

realize

people

due

call

brand

fact

great

started

environment

believe

etc

always

interested

initiatives

API's

Converged

Consumer

social

good

much

usually

time

experience

innovate

NewBay

May 5th, 2010

NewBay is a privately owned Irish company for digital lifestyle services, enabling telecom subscribers to create, store, manage, view, and share user content across screens of PCs, mobiles, TVs, and in-car entertainment systems. With offices in Seattle, Palo Alto, London, and Dusseldorf NewBay is successfully winning clients across the continents and industries - from telecoms like T-Mobile, Telefonica O2, Orange, or AT&T to consumer electronics manufacturers like LG Electronics.

In July 2010 Detecon's Thorsten Claus had the craic with Timo Bauer, Senior Vice President and General Manager for the Americas at NewBay. Timo shared NewBay's vision and insights on one of the largest consumer-facing telecom cloud solutions; and how Cloud is more than just about operational efficiency or cost savings, but can increase a telecom's ARPU, drive messaging and data traffic, strengthen customer loyalty, and build mobile communities and social networks based on user generated content.

THORSTEN: *[Laughter]* It's funny that we always meet in Irish pubs. What a cliché.

TIMO: We also have a CEO named Paddy. But no sheep.

THORSTEN: So why don't you tell us a bit about what NewBay does and what you are currently working on?

TIMO: Sure. NewBay provides a Cloud-based User Content Ecosystem to Service Providers and OEMs. We call our white label platform LifeCache and it enables our customers to offer user generated content solutions like Social Networking, Digital Vaults and Network Address Book services. NewBay has major installations across the world with many Tier 1 carriers and OEMs.

THORSTEN: Hunting Tier 1 telecom elephants is quite time consuming. Did NewBay start out with that vision from the beginning?

TIMO: In the early days we provided end to end solutions to our customers but realized that we need to provide customers with more flexibility and also let 3rd parties innovate against our technology. Driven by this demand we developed an extremely well defined and executed set of APIs. Today many leading design agencies and major OEMs are developing against our APIs. This has been a great success for us and places us at the core of user content systems globally.

We always believed that the telecoms will play a role in consumer Digital Lifestyles. We started in 2002 when carriers did not realize the true potential and strategic opportunity they had to drive their subscribers' digital lifestyle. In the last two years, User Generated Content is finally becoming strategic to carriers and is now part of C-Level discussions. These days carriers are cooperating with but also competing with companies like Google, Apple and Facebook. NewBay enables them to do that.

THORSTEN: That's interesting – these are big companies to compete with, with a strong Cloud vision and strategy. How is that going to play out? How are Cloud markets going to develop?

TIMO: Obviously there are several segments of cloud offerings. Talking about the Digital Lifestyle of the consumer I would like to focus on the consumer cloud. The consumer cloud from NewBay's point of view consists of a Premium Cloud and a User Generated Content Cloud. Whereas the Premium Cloud got most of the attention in the past (Ringtones, Music, etc..), the opportunities surrounding the User Generated Content Cloud are now being recognized due to the media hype around MySpace initially and then YouTube, Facebook and Twitter. Service Providers slowly realize that they cannot just be a dumb pipe. I do not really like this analogy. But in reality, carriers need to differentiate to be competitive and differentiation through voice pricing plans will obviously not last much longer.

NewBay's LifeCache Solutions utilize both cloud types. Even though we are not a content provider or aggregator, we are interested in the metadata around premium content. For example, music playlists, ratings, or the fact that my best friend likes a certain movie. All this metadata around premium content is important for carriers to store and maintain for data analytics, semantic engines etc... There are lots of opportunities for carriers to use that data, such as targeted advertising, social profiling etc. This helps carriers to compete and properly co-operate with companies like Google. It even gives them a huge advantage that they hopefully realize to use. NewBay is working hard with several carriers to make this reality and I think the carrier's are getting more and more excited about this opportunity.

Another obvious example is the Converged Carrier Cloud. Carriers like DT, Orange, AT&T, and Verizon have the opportunity to offer Cloud services across many screens which makes those companies a crucial part of the subscribers' Digital Lifestyle. The fact that I can manage my content seamlessly through my PC, Mobile Phone, Tablet and TV at home is a good customer proposition. Bundled with the trusted brand and proper customer support, carriers can provide a great subscriber offering and experience.

We call this the Converged Consumer Content Cloud, the 4C's. [Laughter] And we believe strongly that carriers as well as OEMs have a great opportunity to be major contributors and even to be the main drivers in this space.

THORSTEN: Carriers get a lot of face-time with on-deck solutions. If something like the Converged Consumer Content Cloud becomes ubiquitous, what impact does that have on how we use or consume Cloud? And what are the most common but yet most commonly unanticipated pitfalls when planning, rolling out, and using Cloud services?

TIMO: I think the consumers are just slowly realizing what the consumer cloud means and really enables for them. Today most people think about how to get content onto certain devices. The "Sync" paradigm is all over the place. We think that the "sync" paradigm is actually "anti the cloud". It is not about "syncing" content rather about enabling easy "Access" to content that sits in the cloud. To me that means that the cloud is not really understood and it may still be a bit early. An example: I have around 10 thousand family photos. I want those to sit in a secure cloud that is automatically backed up. Basically, a safe place. Even today I have actual copies of the photos on many devices and drives due to access limitations and sync offerings.

This will change with the availability of high speed, always on networks that are coming soon.

Again, I think this is a huge opportunity for carriers. I personally am more than happy and even prefer to use a trusted Tier 1 carrier brand to do that for me rather than a Web Service with questionable [terms and conditions]. Who do I trust to manage my personal content for the next 30 years? There are not many companies with the right intentions and capabilities to do that.

THORSTEN: An interesting comment about intentions of companies. What are key success factors for telecoms in a highly competitive ecosystem with many good services to choose from?

TIMO: NewBay started about 8 years ago and it took us a long time and many lessons learned to be able to provide Carrier-grade systems. We are talking geo-redundant five-nines solutions. There are not many companies in the world that can do that.

THORSTEN: Most Web-based consumer-facing companies promise a 99.95% availability in their [terms and conditions]. That's a good eight hours of downtime per year. Some companies also boast with a 99.999% – five-nines – available due to "unplanned" downtime. So who knows about their planned downtime…

TIMO: Good point. But coming back to the "access the cloud" paradigm, it is crucial that the Cloud is available at all times. Downtime is a major issue and as long as the Cloud is not backed by a stable and scalable environment, consumers feel forced to keep their content on several devices. For carriers this is a major challenge as they will get bombarded with customer care calls which obviously increase costs.

THORSTEN: But how can telecoms improve their speed of innovation to compete with companies whose core business are consumer-facing Cloud services? What roles could telecoms play?

TIMO: I would like to give you another example. Let's call it "Cloud Enablers". We believe these areas will be one of the most strategic areas for carriers. Today you see more and more carriers eliminating their "real" unlimited data plans. Obviously heavy smartphone data services are a major challenge for the carrier network infrastructure. One of the biggest issues for the carriers is that they have limited control over how applications interact with the Internet. The constant polling of social networking updates not only causes grief for network capacity planning but also results in a poor user experience due to decreasing battery life. There is plenty of room to optimize data services and I believe the Carrier needs to step up and guide the community.

Our LifeCache Social Networking Gateway for example is used by many carriers and OEMs to provide a social networking service via clients and WAP. It allows third parties to develop services based on our aggregated social networking feeds using our SNG API's. Over the last years we put a lot of effort into our Developer Support Program that provides a sandbox environment and developer resources to use our APIs in the most efficient manner. The initial intent was to provide third parties with flexibility in terms of UI design, but we quickly learned that a huge emphasis is to educate the community on how to use the APIs and create call flows/UI scenarios in the most efficient way to reduce network traffic whilst providing a strong user experience. In fact we have current deployments with major carriers where we provide our SNG as a Multi Service Enabler. The carriers will demand OEMs and other third parties to use our APIs to develop Social Networking Applications. This gives them the control to make sure to optimize all efforts to keep network utilization optimal.

More exciting than this – I consider this exciting...

[Laughter]

... is that this enables the carriers to store all the meta data as the traffic goes through the carrier gateways powered by NewBay. As briefly mentioned before, this presents a huge opportunity for carriers to utilize the data for many services. Obviously, looking at recent Facebook Privacy issues, carriers need to be careful how they play this hand, but I am convinced that they will do it right. This is really how I see the carriers in 10 years from now. You can call this a Smart User Content Ecosystem, or a "Smart Pipe" which it is sometimes referred to in the Industry. I see this as the re-birth of the carriers as a Service Provider.

THORSTEN: But before you said that consumers want the stability and reliability of a carrier brand – five-nines. Now you want to provide enablers that would not be customer-facing anymore. So why are enablers important or even relevant to the consumer?

TIMO: Consider this, today millions of people share content across many sites without even reading and knowing the [terms and conditions]. The Facebook privacy discussion is just the beginning in this regard. So why not provide a technology platform to carriers that allows them to share all the user generated data with third parties, BUT also allows them technically as well as policy driven to retrieve the data back. The carrier will act as your trusted partner in this case and I believe no one in the value chain is better suited to do this better than the carrier. Of course NewBay is more than happy to help and make it happen.

THORSTEN: *[Laughter]* Is there anything – technology, service, product, standards, ... – you wish you would get from a carrier or you could see carriers providing to you? And what do you think are current roadblocks or challenges that we don't see these things yet?

Timo: Well, we are driving the industry to use our API's and make them the de facto standard for the User Content Ecosystem. Today Samsung, LG, Nokia and other major OEMs are developing solutions against our APIs. You have major Tier 1 carriers using our solutions and APIs globally. I give you another example that our LifeCache Social Networking Gateway (SNG) API is basically a de facto standard already. LG for example uses our LifeCache SNG for their Android Social Networking applications. These are the same APIs other OEMs and carriers are using. Already some carriers have asked the OEMs to use the NewBay carrier SNG vs the NewBay OEM SNG so they have better control over the activity and also can start storing the metadata around the Social Networking activity. As the APIs are the same, all the OEMs have to do is to route the API calls and queries towards the carrier SNG Platform. OEMs are not interested in storing or managing data around the usage and optimizing the network. They want to sell hardware …

Thorsten: … for the moment. But they won't shy away from drinking the carrier's pint if they come too late to the party …

Timo: … true, but on the other hand, carriers want to own the Gateways for several reasons. This makes life so much easier for all parties.

There are obviously standards and initiatives we are closely following like LTE, IMS/SIP and other related initiatives. We are network agnostic in what we do though, however we need to make sure our platform supports all standards.

Thorsten: You mentioned LTE – how important is the network? What is "network" in the future of Cloud?

Timo: We talked about Access already. It is critical to have an always-on environment. The network is the backbone for this. So it is a critical component. No network, no Cloud. Simple as that.

In an all-IP-based environment you will see a vast range of IP-enabled devices accessing the Cloud. I am sure we will see more devices that are more like screens and we will see more applications in the Cloud versus locally stored on the device. Traffic patterns and network usage will change. As opposed to sharing actual content, you will just share the reference of the content. Twitter and TinyUrls introduced the concept but we will see it on a way bigger scale providing a better user experience. Companies have started to invest billions in data centers to be ready for this. Again, this is a great opportunity for carriers.

Thorsten: If you say that Twitter and TinyUrls only introduced the concept, how will interactions and transactions with and within Cloud look like in 10 years? Will that be different for emerging carriers and mature carriers?

Timo: I usually differentiate between Tier 1 and Tier 2 carriers. Obviously the subscriber base plays a role here but generally Tier 1's are converged carriers that offer services across major screens. Back in consulting we called it Quadruple Play.

Tier 2 Carriers usually do not have this ability and focus on Wireless only.

To power a Cloud-based platform you need a decent size budget and Tier 2's usually do not have that. So we will probably see different paths for those carriers.

Today we offer solutions to both Tier 1 & Tier 2 carriers. However Tier 1's go in the direction of enabling platforms and let third parties innovate against it. I am not sure how Tier 2's will be able to do that. Perhaps we will see consolidation as well as more partnering in this space to be able to do the same. Either way we provide the technology and add value in both scenarios.

THORSTEN: But what are we going to talk about in 10 to 15 years? What is the next major evolutionary step in Cloud, how will we know that it happened – what's the next "sonic boom" in Cloud?

newbay™
Powering My Digital Life

Billions of media stored

Billions of page impressions per month

TIMO: That is quite far out considering what we talked about 15 years ago. Again there are several segments in the Cloud. The consumer Cloud as defined before already exists and Carriers today are offering Cloud-based solutions. The "sonic boom" though, if even that big, is the burden that people need to understand the access paradigm versus the syncing paradigm. They need to trust the Cloud and that is not quite the case yet, I think. As long as this will happen and we have a solid network and Cloud-based infrastructure, the consumer Cloud will dominate and set the new standard for how we manage our content on a daily basis.

We invested significant amounts of time and resources in system and software engineering to develop platforms that provide and enable those requirements. In fact we are going to launch the biggest Converged Consumer Cloud with [a Tier 1 carrier] later this year.

THORSTEN: it's going to be interesting to see how fast the consumer mindset will change with [a major Tier 1 carrier], having access to so many consumers across so many screens and digital lifestyle properties. Thank you for sharing your thoughts and insights.

TIMO: Thanks so much for inviting me and I hope I gave you a good perspective how we think the consumer content cloud will develop and the importance of the carriers.

running two first lot unreasonable global high quickly

made software availability reason accounts obviously ever used

environment complexity technology tools find done around customers number years

service make community sell entire postpone brand capabilities code

shared data relevant management idea also get

provider enables hours user currently never security amazon

function capability answer chaos force.com like actually time process

solutions relationship means already thriving cost whole model many

cloud customer system new one important instead

companies go else's perform

every hear infrastructure called demand year salesforce applications vendor

able different benefit developers java salesforce.com check part

core fundamental wants forward achieve

iphone believe resources build run enterprise

business world virtualization someone something microsoft offer always computing way story rapid

problem just value intensive

support marc crm

market call now people use information going thing centers

need capital company last architected today write

Salesforce
October 30th, 2009

Ten years ago, Salesforce.com Inc. was founded by former Oracle executive Marc Benioff as a pioneer in delivering enterprise Software-as-a-Service. Their Customer Relationship Management (CRM) product was a huge success; after several years of development, its radically extended and enhanced foundation was exposed to developers as the Force.com Platform, allowing external developers to create CRM add-ons and also to build completely new applications that remain hosted on salesforce.com's multi-tenant infrastructure.

Launched in 2005, AppExchange is a directory of applications offered to the salesforce.com marketplace by third-party developers: users can subscribe to these offerings and integrate them into their individual organizations' salesforce.com environments. There are over 800 applications available from more than 450 Independent Software Vendors (ISVs).

We caught up with salesforce.com's Director of Platform Research, Peter Coffee, in a conversation with Detecon's Thorsten Claus to learn more about the experiences driving these decisions and where the company plans to go next in its relentless efforts to continue growing through a down economy.

THORSTEN: In 2000 we had a lot Application Service Providers (ASPs). Then everyone moved to "On Demand", with a lot of people having a lot of different opinions what that actually means. Now everyone is talking about "Cloud". What is different from 2000?

PETER: The fundamental difference is that application service provisioning took the load of configuring and administering a copy of an application for a customer out of the customer's hands and allowed the customer merely to use the function, but that workload didn't go away – it was just assumed by the service provider. The complexity remained. The difficulties of patching and updating software remained. The difficulties of coordinating a complex and brittle stack of software products remained. You weren't able to get massive economy of scale or dramatic acceleration of the rate of technology improvement in that model.

In the Cloud Model, we've attacked the fundamental obstacles to economy and efficiency in enterprise software by building architectures that were conceived from the outset to embody massive sharing of the parts of the system that don't require intense customization.

For example, everybody wants the best security you can get. There is no good reason why security and trust components of a system shouldn't be architected into a massively shared foundation so that all security improvements benefit all users of the system.

At the same time, using a metadata architecture for customizable parts of the system enables the same kind of rich, deep customization and automation and tailoring of business processes that people have always required – and will continue to require – of enterprise software solutions.

What that means now is that when a cloud computing provider like salesforce.com improves core capability – like query performance, user interface customizability, addition of new features to improve security and the management of privileges –

all 63,000 customers simultaneously benefit from this, and all of their customizations are segregated at the metadata layer so that they move forward without disruption.

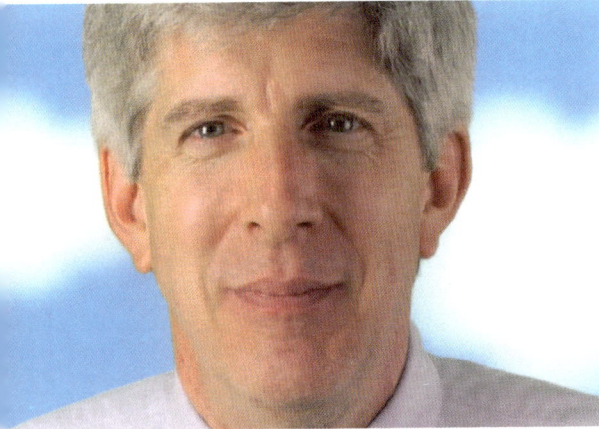

And that is the core distinction between the old ASP model that simply relocated complexity out of the customer's basement and into the service provider's basement: In the old model, because it was so disruptive to perform a major software upgrade, customers would tend to postpone those upgrades and sometimes even skip an entire release of a product as has happened in the desktop environment over the last 5 or 6 years.

This is bad for two reasons: One, it means customers aren't getting the benefit of new technology because they postpone the acceptance of an upgrade; two, the vendor now has to support and provide security patches and other issues have to be addressed for many versions of their technologies simultaneously. The vendor's efforts are diluted while the customers' benefits are delayed.

In the new model, a Cloud vendor is always maintaining a single, coherent codebase which can even support legacy behaviors if the vendor decides to do so – as salesforce.com does. We don't force customers to accept new behaviors, we offer them new behavior to enable or not as they choose. But the code base is coherent, so we can improve at a more rapid rate and customers have fewer reasons to postpone the acceptance of new function. Everybody winds up moving forward more quickly with less risk at lower cost.

THORSTEN: So the difference is a novel architecture to combine massive economies of scale while retaining rich customizability in ways that don't just relocate complexity but substantially collapse complexity. In many industries this important message about the Cloud Model often gets confused and **reduced** into something about cloud computing – what you hear over and over again is a story about virtualization and distributed data centers and Amazon S3 …

PETER: … Yeah, I hear that, too. But that's only half the story. Virtualization is a very important enabling technology. But nobody gets up in the morning and says, *"gosh, my business problems would be solved if I could only go buy some virtualization."* What people say is, *"I wish that I could innovate more quickly,"* or *"I wish that my IT costs were more predictable,"* or *"I wish that I could deploy new capability without capital investment."* And those are the issues that I believe cloud computing fundamentally addresses. Virtualization is one of the important tools that makes that feasible, but that's a supply side fascination – it's not why the demand side is actually interested in this.

THORSTEN: Now salesforce.com has always existed in the Cloud – one of your trademarks is "No Software".

PETER: But it's more than that. Salesforce.com came into being when Marc Benioff said, *"why is it that consumers have capabilities like iTunes, Google, or personal recommendation engines on Amazon that they can find and learn and use and that continually improve, while enterprises cannot get rapid innovation and must tolerate unpleasant costs and unreasonable risks in their technology?"* Marc strongly felt that there is something fundamentally wrong with that and set out to take a specific category – customer relationship management – where any business could benefit. Every business from the small high growth business to the mature organization is looking to get more value out of its existing customer relationships. Marc believed that there's no reason why enterprise capability should not be something you can combine with discoverability and ease of use. That goal and vision was the birth of the company.

We've always been a pure cloud player – we just were not always called by that label, which really only emerged about two years ago from the concept that no one really wants to own infrastructure except the people who sell infrastructure capacity. Do companies operate their own fleets of delivery trucks? No, they call UPS or DHL or FedEx to get delivery done. There is no reason why IT shouldn't predominantly be done the same way.

THORSTEN: But isn't that idea called "On Demand"?

PETER: There have always been technology vendors who have found that an "on demand" service model is a way to sell technology that's more palatable to the customer in times like these when capital is limited – but they are still fundamentally selling you technology. We have never sold software, technology. We have never delivered a piece of technology to any customer, ever. All we sell is service. We only build technology to enable service. And that means that our focus is on creating and nurturing a relationship with the customer in a way that is fundamentally different from the model of any company whose DNA is being a technology vendor.

THORSTEN: Talking about the service focus – Salesforce also enables companies to have access to a whole ecosystem of developers. Do you think that it's the common platform and processes or the access to your customers that are attracting developers to Salesforce?

PETER: There are more than 450 companies offering more than 800 different applications that all directly interact with the salesforce.com platform. That number is growing enormously now that we have exposed the foundation of our application suite to the community as well, what we call the Force.com platform. It enables the construction of an application running entirely within our systems, or integrated with other resources either in the cloud or in local data centers.

Developers can come to us with nothing more than an idea and something that runs a web browser, and build an application; bring that application to market under their own brand; and immediately be able to offer that application to a global community.

Two guys in Bangalore with laptops could credibly be providing technology to the global 2000 companies in two months if they have an original vision of how to solve a business problem and build it on the Force.com platform. They need no capital to build a server farm, nor do they need to prove their ability to administer security because they're running in an environment that's already been audited -- nor do they have to prove high availability and high up time, nor do they have to field a global sales force.

You can have a rapid entry of talented entrepreneurs with almost zero capital, and with low risk on behalf of the customers evaluating these new market entrants.

THORSTEN: That sounds great, but is it possible to build a sustainable business model based on someone else's interests?

PETER: First of all, there's a new class of boutique integration firms, companies that are building completely serverless IT portfolios for companies. One of my favorite stories is when the the CRC Health organization acquired another health care provider and discovered that it was now dealing with a very heterogeneous IT environment that was both complex and not very scalable. So they used the Force.com platform to build a highly integrated system that incorporated the things they had that they wanted to continue using, and at the same time gave them a new system with a shared view of key data across all their various points of presence.

I like that story because people sometimes suggest that the Cloud is not suitable for highly sensitive, private information like medical records. They suggest that compliance with regulatory environments will be difficult to achieve. I'm not saying it's easy, but compliance is absolutely achievable in the Cloud.

THORSTEN: But many of our clients are not only concerned with security and privacy but also the reliability and availability of their data. While many of them have an expensive double-redundant backup of their local data to geographically different backup centers they are concerned whether a Cloud provider could achieve the same reliability.

PETER: True, but let's compare to what you currently actually spend and what you currently actually achieve in terms of availability. In our most recent quarter we achieved 99.997% of planned up-time, and our scheduled maintenance windows are continually shrinking. Meanwhile, you'll discover the scheduled down time for maintenance of your average email system in a month is typically greater than the entire downtime – planned or unplanned – of a competently run Cloud based system over an entire year.

THORSTEN: If you offer all of these features easily accessible to a global audience of developers, how do you make sure that the developers don't harm the brand of salesforce.com by doing something that's not allowed, unsafe, or violate operational conditions? What happens if they found the killer application and basically bring down your system because there are so many requests you suddenly have to handle? And what if that killer application happens to be your planned future core business?

PETER: You put your finger on the key thing: How do we make it possible for someone to create independent intellectual property while running it in a shared environment. The only cost to the developer is a transparent fee we pay a third party provider of security services to validate the basic security of an app and ensure that it's not doing anything inappropriate that would create pathways into our platform that would be hazardous to anything else that's going on. Not only is this audit function vital to us, it's so vital that we have it done by an independent party so that no one will ever have to wonder if our zeal for growth is making us do something careless.

The second thing is that our platform has been architected with the understanding that people were going to be creating custom logic that has to run in a shared environment. People ask us why we created our own programming language, which we called Force.com code. If it looks just like Java and it works just like Java, why didn't we simply create a facility in which developers could run Java in our environment? The answer is this: If we had taken the standard off-the-shelf Java Virtual Machine and embedded that in the Force.com environment, we would then have found ourselves forced to build walls around that to prevent it from ever degrading the shared facility. If someone did something incredibly processor intensive, we would have to add all sorts of protective mechanisms around it. So instead, we implemented a Java family language in the multi-tenant environment so that if you try to do something that's going to place unreasonable burdens on shared resources, the system is able to give you a well-behaved exception mechanism that says, "I'm sorry, I can't let you do that."

It wouldn't be acceptable if all it did was abort unacceptable behavior. We implement capabilities where a well-constructed application can query the platform on functions it wants to perform. Returning a million results from a user's poorly formed query is obviously not useful to the user, and obviously it's not an

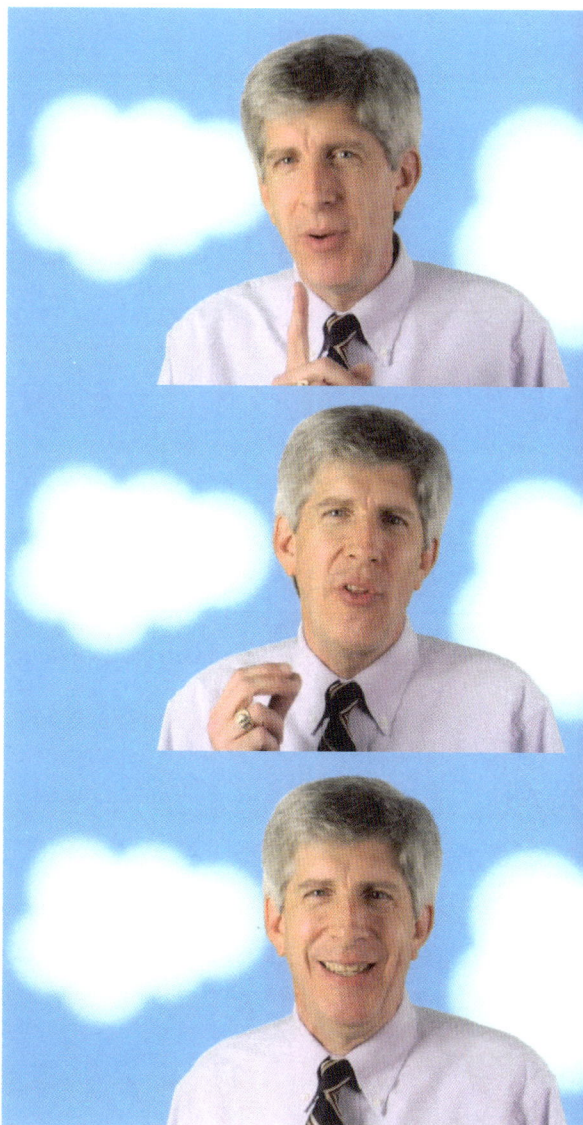

acceptable burden on the platform either. So you can write applications that provide a well-behaved user experience while at the same time respecting the need to avoid unreasonable burdens on shared resources. As a result, we're able to make very rigorous statements about preserving performance, maintaining data integrity, and assuring that none of the applications running in our platform becomes in any way a pathway to invidious acts that would damage the experience of other applications.

THORSTEN: People love your story about respecting shared resources. But in a world with seemingly unlimited computing infrastructure of Amazon S3, a Sun cloud, and Microsoft Azure at my fingertips, why bother? How do you compete against the rapidly emerging CRM suites built on this cloud infrastructure such as NetSuite, ZohoCRM, Sugar CRM, Heap, and Microsoft Dynamics CRM? How do you prevent that Salesforce CRM becomes a commodity?

PETER: What I always emphasize is that the Cloud is a multi-vendor, multi-product marketplace. In the early days there was a perception that every Cloud service was its own little world and that you had to find the Cloud service you are going to adopt to the exclusion of others. Instead, we always architected for integration with other solutions, so no one ever had to feel that adopting us would prevent them from using the best complementary solutions that were available from others.

In fact, most of our customers are probably using us in combination with other facilities. They may be using us to pull key information in from an SAP or Oracle system, and then make that information relevant to people in the field who they could never serve with current data before. They may be using Amazon facilities so that they can spin up virtual machines in Amazon, perform very compute intensive or very storage intensive tasks more cost effectively while at the same time, using our facilities, build applications much more quickly.

We now have four separate studies that all agree that on average, developers will produce the same amount of capability by using the Force.com environment – where many complicated behaviors are now configurable services instead of requiring code – and will be five times as productive. They will write something in 20 hours instead of 100 hours compared to developing the same capability in Java or .Net.

The study that I like the best was done by a company whose product is software cost estimation software – tools for doing software project management. Their only product is the credibility of their cost assessment. They have absolutely no reasons to bias the number in our favor because if the estimates they produced for us turn out to be wildly optimistic, their own brand is damaged.

THORSTEN: What I was relating to was the fact that many companies copy with pride salesforce.com's lessons-learned and user-experience up to the interface and start addressing the same target market, minus the AppExchange and platform approach. Is that part of salesforce.com becoming a commodity?

PETER: Here's my answer to that: Every time another company gets into the business of selling its expertise in the form of an application that you use as a service, I have one more set of voices out there helping me to make the customer comfortable with the idea that they can host sensitive data in someone else's system and run important business logic in someone else's system. The most expensive part of our process of acquiring a new customer is simply getting them comfortable with this whole Cloud model in the first place.

So every time I get a great, big gun like Microsoft or SAP or Oracle finally agreeing that the Cloud model is fundamentally a valid model, that's good for me because now I compete on nothing but the pure business value and demonstrated care for the customer that make a top tier service provider, which is different from being a top tier technology provider.

THORSTEN: So what's next for salesforce.com? What do you see on the horizon? What's the next big thing that you're working on?

PETER: We're very careful about forward looking statements. We're a publicly traded company, and so I can't say anything that isn't already rooted in something that we're doing today.

But if you looked at what we've done in the last three or four months, we've taken functions like service and support and instead of merely providing the same capability as the world leading service and support tools, but in the Cloud, we thought about how to re-conceive the fundamental mission of a service and support tool given the existence of the Cloud.

Because customers today no longer pick up the phone and call your toll free number and call you as step one of a problem-resolution process. The first thing they do is go online, check Google, check their community on Facebook, check Twitter, check any number of other resources where they think they can find a credible, independent expert solution … and maybe, as the last resort, they call you.

We wanted to wrap around that process and built what we called the Service Cloud, where things like Tweet streams and Facebook conversations and other social media are directly incorporated and automatically mined for relevant conversations of which you need to be aware. The solutions that your community is creating for you can be incorporated into your own frequently asked questions and tools. You can be affirmatively engaged with that community instead of being someone that they talk about, with you not in the room. And it's a far stronger post-sale relationship with the customer. This is of immense value to any business.

You can see this in what we've already done with Service Cloud, which was deployed recently -- and obviously, as social media become more and more sophisticated and more and more embedded in our lives, it's only to be expected that our Service Cloud will be evolving to stay abreast of that.

THORSTEN: Will mobile be part of that ecosystem?

PETER: Yes, definitely. We had always conceived of the client as merely a thin interaction layer. We didn't need to be able to write specific SDK code that ran on the iPhone. We have continued to track the emergence of new developer capabilities of the iPhone, certainly, but the amazing statement was that we ran on the iPhone on day one.

THORSTEN: The easy-to-roll-out mobile experience of salesforce.com was a decisive factor for many enterprises to tap real-time information, especially for the brand management aspect of not only their direct customers, but also their users – which might be completely different, as you mentioned before with the application development opportunities …

PETER: … exactly. I believe the Service Cloud is probably the exemplar of turning that from a problem into an opportunity. You will hear us talk about the real time Cloud because now, with the necessity for rapid recognition and instantaneous response, you can't have systems where time lags of hours or days or weeks or months are part of the way you do business.

You have to be able to say, the minute we know something, subject to appropriate security reviews, we can immediately surface that information on our website. That's what we did the late last year with the Force.com Sites capability, where now you can have a public website: not something that requires log-ins and authorizations, but a publicly facing website that has a direct, genuinely seamless connection with data that used to be internal.

It used to be you developed some new content, then you had to call up the guys in your Web site production department and say, how long will it be before you can get this up on the site? And now the answer is: hit the Refresh button, it's there.

THORSTEN: But you state an interesting – and difficult – problem: "Real time" obviously also means, how do I filter all the irrelevant information with the relevant information?

PETER: Yes, the world is awash in data and now we have to figure out how to how to turn that into actual information.

THORSTEN: But it's easy to say that you need to filter relevant from irrelevant data, connect the dots, and present information, than to actually do it.

PETER: Well, that's where we made major investments in the last year. We made acquisitions in the areas of a data mining. We made enormous improvement to our organic analytics capabilities. We have very strong partner relationships with companies like Informatica and Pervasive and Cast Iron and Tibco whose whole expertise is in integrating data, repurposing data, and finding ways to get more business value out of existing data.

Ultimately, it all comes down to your relationship with your accounts, with your contacts at your accounts, and with the decision makers in those accounts. Oddly enough it all seems as if everything is in some way rooted in what we used to call CRM because what it comes down to: what's going to happen, what actions are going to result from this information. A person is going to make a decision.

THORSTEN: Theodore Roosevelt once said that the most important, single ingredient in the formula of success is knowing how to get along with people.

PETER: And one of my all time favorite business books is Tom Peters' "Thriving on Chaos" that he wrote about 20 years ago. He didn't title it "Thriving in Spite of Chaos" or "Thriving in Defense Against Chaos". He titled it "Thriving on Chaos", recognizing that chaos – by which he doesn't mean confusion, necessarily, but constant, turbulent, unpredictable change – is a defining characteristic of the business environment, the government environment, and the healthcare environment; and that you should make your ability to deal with that chaos, and find value in it, one of your defining competencies. That is what will make your success: when your customers are better able to cope with this environment because they're working with you.

It's hard to have a sustainable business model of being the smartest guy on the planet, forever. But you can always try to be the partner of first resort for whoever is currently the smartest guy on the planet.

I believe that today, if you have an idea that you believe you can bring to market in the form of something that some that people will find valuable, you need an email address, a bank account and a Force.com account: you are visible, discoverable, and you are able to make the financial transactions that turn you from hobby into a business, and you are able to package what you know in a form that other people could use and add value on top of that. And I believe that that's how businesses today and tomorrow are going to be created and will succeed.

THORSTEN: Well, I'm looking forward to see any payment and micro-payment platform from Salesforce soon – hopefully mobile, too – and then I don't even need the bank account anymore.

PETER: [silence] What an interesting idea.

THORSTEN: Thank you for your time.

PETER: My pleasure.

companies
think
just
infrastructure
cloud
business
get
lot
going
want
basically
actually
service
services
really
today
technology
software
operator
people
large
now
see
space
network
applications
technologies
cost
time
one
kind
application
company
operators
end
stuff
investments
industry
problem
different
virtualization
interface
things
like
point
capital
revenue
sure
go
know
getting
Service
back
bandwidth
buy
way
always
early
faster
Yeah
easy
data
play
together
another
startups
bunch
pricing
venture
consumer
makes
mean
probably
make
anymore
maybe
still
year
first
offer
et
TV
sales
limiting
fiber
big
world
stage
channels
put
web
hardware
perspective
usually
years
carrier
telecom
video
side
majority
something
energy
carriers
change
look
looking
need
making
U.S
stored
bit
centric
much
funding
cetera
whatever
facebook
enterprise
air
interesting
may
pretty
invest
difference
two
experience
scale
right
vc
advantage
model
Interesting
next
sell
factor
around
build
innovation
good
order
little
pay
devices
question
many
Oh got
tend
done
cycle

Sofinnova

December 1st, 2009

Since 1974, Sofinnova Ventures has partnered with entrepreneurs to secure initial funding, build successful teams, win key customers, and navigate acquisitions and IPOs. Sofinnova invests in early stage information technology and life science companies in Europe and the U.S., typically as early stage, series A, first institutional investors. Their current technology portfolio includes companies such as Openwave, UPEK, Laszlo, HelloSoft, Crocus Technologies, or Streamezzo.

Detecon's Thorsten Claus caught up with Brian Wilcove, Partner at Sofinnova Ventures, to discuss the venture capital perspective on Cloud computing, as well as fascinating insights on future areas for Silicon Valley investment and telecom opportunities. The following article is a summary of some of the highlights from their discussion.

THORSTEN: Can you talk a little bit about what Sofinnova does and about your position?

BRIAN: Of course. We're a venture capital firm. We've been around since 1972. We invest in early stage information technology, IT, and life science companies. Typically, we are early stage, series A, first institutional investor in companies.

THORSTEN: A lot of my clients ask me *"what does Cloud do for me?"* – What does Cloud do for your clients or what do you think it could do? Or for yourself? Where do you see Cloud growing?

BRIAN: As a VC, it's a segment where there's a lot of investment. And I'm assuming they'll be some exits in the space because people see it as a hot space. Beyond that I'm not quite sure.

THORSTEN: The way you phrased that it seems very different from what you would tell me from a personal experience.

BRIAN: Well, from a personal experience: do I want access to my photos in the Cloud, video on the Cloud – basically not stored on my local PC so I can keep track of it there? Sure. Do I want my contacts stored in a Cloud? Sure. As a person: yes. As a business, I have no idea what it actually means other than a difference in how people sell software.

THORSTEN: Interesting. So you don't see any differentiation between cloud and on-demand and Software-as-a-Service (SaaS)? Is it just another term?

BRIAN: It's just another term.

THORSTEN: So is it just something to confuse the telecom, to sell more stuff, or different things?

BRIAN: No. I don't think telecoms are confused. This is what's bizarre. Before venture capital I worked in a variety of telecoms. And one of the businesses I ran, we called it "ASP [Application Service Provider] business", which was basically hosted applications. And that, basically, is the Cloud business today. The only difference being that back then the software architecture was client-server. So it was more difficult to integrate multiple applications together. And now, to me, the only difference in Cloud is that it's all a web-based protocol with easy integration to back ends. But it's all the same stuff. It's just a business model.

THORSTEN: What's the difference in this business model to the business of an Application Service Provider?

BRIAN: We've gone through a lot of industry cycles over the past several decades. We started off with large mainframes where you would basically have to timeshare yourself, have punch cards, you have a schedule, et cetera. And that was a Cloud. And then it got so cheap that you could build PC's and have every person having one, or now, multiple netbooks, laptops, devices in your house. That became ubiquitous. But the problem was: because of the ubiquity the cost of management went up.

So now we're back to the point of: *"The cost of management is too high. Let me just put it all in a centralized location because bandwidth is basically a lot cheaper and more accessible today."* We're kind of back to the mainframe business. That's what Cloud is. So it's a business model. It's just one way of a service provider, and that could be a service provider in the largest sense, not just a telecom, getting economies of scale by aggregating computer resources together.

THORSTEN: You said that there are going to be a lot of exits in this industry. From a VC perspective it's a hot sector right now.

BRIAN: Yeah.

THORSTEN: What are you looking for in companies?

BRIAN: We made one investment in a company called ConteXtream. They do network virtualization. So the concept there is basically that there's a lot of transport and routing and switching infrastructure in a network telecom that has its own cycle of evolution – technology cycle, depreciation schedule, et cetera. And then there's the application domain, which is changing much more rapidly than the underlying infrastructure. Today the way applications get built in networks is that they're very much tied to a specific set of hardware, software, et cetera.

In the case of ConteXstream, they build software that sits on a compute resource, like a server, and segregates the application domain from the network domain. So it's network virtualization. They do load balancing, global load balancing, traffic steering, these sorts of things. So that's kind of virtualization, kind of Cloud computing, but it's for the network telecom's infrastructure itself. Not as a service to sell. That's one area that we've invested in.

In terms of enterprise Cloud, which is where the majority of the activity is, we haven't done any investments in that space. I don't think there's many good opportunities unless you're really willing to make a lot of investments.

The majority of what I would call "Cloud deals" are enterprise centric. We've seen a lot of SSD, a lot of storage devices, virtualization software, desktop virtualization. It's more around targeting towards the enterprise IT guy than anything else. This migration into using the Cloud, the service provider providing Cloud, there's not a whole lot of opportunity for venture investments.

THORSTEN: Interesting. As a VC, you often have an advisory role and connect people in the industry – with the intent to sell or exit or whatever you want to do with your company. Is there something you would say, *"Oh, I wish I would have that."*

BRIAN: As it relates to Service Provider Cloud infrastructure? I don't think so. I mean: it's not a space I've been spending a lot of time in it. Because, to me, if a telecom wants to play in the Cloud space, it's very easy to go buy a bunch of servers, a bunch of storage from Hitachi or a big name vendor, right? EMC, NetApp, whoever, put it in the Cloud. Buy some VMware on top, virtualization software and start selling stuff. Is there a place for early stage venture-backed company there? I don't think so. It's mostly integration and buying big hardware. It's a financial game.

THORSTEN: Given the fact that so many Cloud computing facilities are out there, isn't the amount of IT innovation much faster now? Are many of the companies you're looking at using Cloud internally?

BRIAN: Yeah, I think particularly, in the application world that is the one advantage startups have now. And a lot of other VC's will claim now that they benefit. A lot of startups we see that are developing applications use EC2 or Microsoft or whatever it is on the back end. And it just lowers their cost of operations, which makes it easier to develop applications, which is kind of partial to this whole web 2.0 thing. We are not investors in that thing.

There are definitely winners to be had in that space. There are many companies that are good companies, just like in the first wave, that will exit with reasonable businesses. But the vast majority will not. Cloud actually lowered the barrier of entry for companies getting into this space. So we actually see, one company will come in and have some business plan. And that same business plan we'll see over and over and over again because it's just so easy to replicate. There's just no barrier to entry.

THORSTEN: That obviously also makes it difficult from a VC perspective because you have a great idea and anyone can copy it.

BRIAN: But I do think from an telecom's perspective that is exactly the opportunity: They can build infrastructure that enable corporations, enterprises, their end customers, consumers to buy and consume those applications much more readily or easily. If I was a Verizon or a Deutsche Telekom or a Vodacom or whoever, I would just go to an Oracle and buy a bunch of stuff, buy a bunch of professional services from IBM, and put a big Cloud together and offer a very similar – there's nothing to stop them from offering a similar service. Because it's all about deploying capital, managing a large operation, and – over time – getting a recurring revenue business. And that's what the Cloud is. Service providers have done that for the past 50 years. There's no difference. It's a big opportunity for them.

THORSTEN: So why do you think telecoms are so hesitant to do exactly that?

BRIAN: I just think that they just move very slowly. They're just large warships in the water. Will they do it? Yes. I actually think that many of them will do it. It's just a question of time. You and I, we live in the Valley. We get ahead of ourselves in terms of the rest of the world's adoption of technology. And we see all these things. And what happened two months ago is old news. To the rest of the world that's not necessarily the case. So I do think that telecoms will adopt it. I do think if you look at Cisco, IBM, Oracle, SAP: All these vendors have very large programs. And that's what they're geared toward: They're geared toward selling to traditional large enterprises for the Chevrolet's of the world and telecoms considering; we will see adoption in that space.

THORSTEN: Many startups in the web business are pushing their often Cloud-based services. You said yourself a lot of them are dying again. Let's say a large company actually does adopt such a service. From your experience, does it actually scale, or is the Cloud in the end a limiting factor? Because you start deploying your service so fast for a small pilot group, but then suddenly you need to scale from 200 to 200 million people – is that a problem?

BRIAN: Telecoms don't do anything without scale. That's the advantage that they have I think they should play on. When I go back to the days when I was running services or products inside telecom operations, I would take the approach of starting with some very basic services. It doesn't have to be sophisticated. There's too many buzz words and too many sexy technologies out there, and it's just hard to swallow. I would just start some basic stuff. Backup services: people have file servers, a bunch of branch offices, why don't you just back that stuff up? And it's not complex. And will H&R Block pay an telecom $100 a month? I don't know. Probably. And over time that becomes a pretty large business. But to me it's just tackling basic services that a telecom ought to offer.

THORSTEN: But maybe startups don't pay that much attention about software design anymore because services are so easy to do. And then suddenly it doesn't scale because their design doesn't scale. Twitter didn't scale. So the question is: are Cloud-based services actually dangerous for you as a VC?

BRIAN: I'm not sure it makes it any more dangerous or not. Because most of the software companies I've invested in at one point or another always rewrite the architecture. They never get it right the very first time. The advantage of having a Cloud infrastructure is that they can get it off the ground fairly quickly. Actually, I think the biggest limiting factor of Cloud is cost. So if you really do the math on EC2 or something, it gets pretty expensive pretty quick the minute you get out of play land. More than the technology scaling there's a cost point where it makes more sense to actually bring it back in house.

THORSTEN: You can have a Network Attached Storage (NAS) for $5,000.

BRIAN: Exactly. So that threshold is actually pretty low today. Where I see traction with Cloud is in startups that tend to be the very early stage where they want to prototype something or get a few thousand beta users or whatever. And then they get funding or a series B comes in and they say, *"Okay, I'm going to bring this in house and have my own,"* in order to get costs down.

THORSTEN: Very interesting. Wouldn't that be the ideal situation for a telecom to say: *"I provide my own internal departments a Cloud facility that they can play around in"*? And as soon as they have an application where they see it's going to work, they get kind of a round B funding – corporate funding internally –, another $2M or $3M for the people to keep them running. Then they say, *"Okay, now I push that service to something bigger, better."*

BRIAN: Yeah, you're asking a question of whether an telecom can fill that void, basically become the next kind of step function in the economic value chain – I think that's probably true. Certainly, I think that's probably the place to attack the marketing. I'd go after the super low end.

THORSTEN: Interesting. So what kind of companies are you personally investing in right now?

BRIAN: I'm investing in two general areas. One is infrastructure technologies, that's largely in communications, enterprise infrastructure, telecom kinds of things; and then energy technologies. Those two things.

THORSTEN: Why is that?

BRIAN: If you do an analysis of returns, it's still by and large mostly infrastructure technology companies that are having the big outcomes. So even if you look at the last quarter, you have LifeSize for $400M. You have Starent for $3B. All of the large exits are still actually infrastructure companies.

THORSTEN: Isn't that like playing the lottery?

BRIAN: Oh, no. I don't think so. As a percentage of dollars into that space, actually it's a lot lower. Most people have avoided that sector for at least the last half a decade. The capital requirements are high enough to where only the really best companies in that segment end up getting financed. It's still venture capital. There's still risk. Certainly companies die. But I don't think it's playing the lottery. I actually think web 2.0 is playing the lottery, which is why we don't invest there.

So that's one space. And the other space is energy. And it's not so much clean tech but really the intersection of IT and energy. So I've done a lithium-ion battery company. And I'm looking at a lot of other similar kinds of component technologies for consumer devices, primarily.

THORSTEN: When you invest in those companies, is there a general advice you can give them for each of these sectors? Something you run into over and over again?

BRIAN: Telecom related or oriented companies always have the same problems of sales cycle. This is particularly problematic in the U.S. You're either selling to the large Telecoms, Verizon, AT&T, Comcast, Time Warner, et cetera, in which the sales cycle can be 18 months or longer. And you have to first fund the development of the company. And then you're basically funding this potentially never ending process of sales, which consumes a lot of capital.

My advice to companies that are in that space – and I have several – is usually to partner, and partner early. To defray sales costs and to try to shorten that sales cycle. So partnering with and Ericcson or and Alcatel or a Cisco, whoever it may be. Because by and large the telecom equipment manufacturers are turning into more and more service oriented companies.

You have HP and Oracle on a traditional server and compute side trying to get into that business. You have Ericcson and the same basically wanting to do more and more operations. Selling into those channels is becoming more and more important.

In the energy space it's a bit more pioneering. The investments we tend to make there tend to be more fundamental technology bets: if the technology works then there is an absolutely huge market at the end of it. That is usually around making sure you can really understand the physics, the electro-chemistry of the problem and making sure you have a team in order to solve the problems.

THORSTEN: But aren't both of these markets massively fueled by Cloud?

BRIAN: How do you mean?

THORSTEN: Your network infrastructure investments are targeting more efficient routing, handling large amounts of data, or customizing communication. With more and more mobile devices, anything done in the Cloud leads to a lot more data over the air channel. On the energy side, the always-on connectivity of Cloud requires that I run my cell phone longer or my netbook longer. With your investments my netbook can actually be online all the time and also have a 3G connection all the time without sucking the battery after two hours. Aren't you happy about Cloud then?

BRIAN: Yeah, but so what is the network? Isn't my network a Cloud?

THORSTEN: No, because it's centrally managed. It's not a multi-tenant entity. It's only one tenant and that's called AT&T or Verizon. AT&T provides the infrastructure. While the network, itself, yet has different end points, I can't manage that. There is no service portal where I could say, *"Oh, now I want to resell part of that T1 to someone else."* There's no automatic provisioning. There's no self service. There's often no up flexible upscaling or downscaling, like Yipes / Reliance Globalcom is doing.

BRIAN: So certainly there's bandwidth on demand. So there are examples of being able to sell downstream. So multi-tenant building where you bring in DS3, or fiber or VDSL. You put the switch in …

THORSTEN: … housing developers then basically resell them again …

BRIAN: … exactly. A network is a Cloud. There's not much difference to me: the Cloud these days is basically the fact that it's more application centric, storage centric. It's the fact that applications have web calls as the interface to one another so you can mix data. Facebook is a good example. Facebook is a platform because you can run an application within it. That's a Cloud, I guess, of sorts. But I don't think there's some fundamental magic change and all the sudden VMware got spun out. And everybody put VM's on machines and now there's a Cloud. It's always been there. No change.

THORSTEN: Looking ahead about 15 years – and I say that with a very specific thought in mind – if all of your visions come true, what would we talk about in 15 years from now?

BRIAN: People talk about the Internet as the platform for every kind of communication, entertainment, et cetera. That'll be more and more real over time. We're going to turn in the age where more and more of your data is stored in the Cloud, the server somewhere in the network, and accessible anywhere. Today I can't really easily move music from my home to my car, to my work, et cetera. These things that are visionary now are going to be completely seamless. All my stuff will move with me anywhere I want it.

I'm a big believer in virtual reality and three dimensional projection, holographic displays. We'll see that, in terms of video. Which, to me as an telecom, I'm a little bit happy and a little bit scared. Because that will consumer a lot of bandwidth. And today I'm not getting paid anymore for the incremental bandwidth. That could be another 100 fold increase, if you really think about holographic projection TV. And I think those technologies will be fairly readily available.

THORSTEN: There are two large telecoms here in the U.S. who are starting 3D trials – from 2D+ formats to real 3D. But then again, all this data needs to be stored. And you just talked about conversion between different devices – isn't Digital Rights

Management (DMR) at this point a real stumbling block because you don't want – as a content owner – that your assets get converted to all these other formats where you don't know what the quality might be, or how protected your asset is?

BRIAN: I think that's a problem only today. 15 years from now it gets solved. The problem that this specific industry is having is that the rules are changing faster than the industry can change. And it just needs to adjust. It can try and apply brakes, as an industry, but it only can do that so long to buy itself time to figure out how to basically remake itself. That industry has plenty of smart people on, the content industry will figure it out. There's no stopping what a consumer really wants at the end of the day.

THORSTEN: What kind of infrastructure do we need in order to provide this picture, this seamless connectivity all around me? What kind of infrastructure would you like to see?

BRIAN: The limiting factor and what makes that possible is the air interface technology of mobile communication. I think on the fixed side with FiOS and other kinds of fiber technologies, there's plenty of capacity. I think the limiting factor is really your air interface. Because the vast majority of times you want to consume your data you may not be at home. To have a truly seamless experience where you're not degrading some quality because you're in your car or walking down the street – a great air interface technology is critical, as is the subsequent aggregation and personalized transport.

THORSTEN: ConteXtream has the interesting approach where a telecom can virtualize how antivirus filtering IT, whatever you have, can be offloaded into a Cloud, making it very attractive for telecoms to deploy several SLA's. Is that part of the solution?

BRIAN: The intention of ConteXtream is to make life a little simpler and enable the telecom to innovate faster on the application world and decouple from the underlying transporter infrastructure. And I do think that that's going to be necessary. Because Facebook is really popular today. Two years ago MySpace was popular. Twitter is popular today. We just don't know the pace of innovation and adoption …

THORSTEN: … what's the next Twitter, what's the next Facebook …

BRIAN: Yeah. So for an telecom, they need to have a way to introduce applications in a pretty rapid pace and actually get fairly granular. Today, I get a phone service that's the same for everybody. It's universal, but it's probably not really optimal. Optimally, what you want is a very targeted set of applications, pricing, network access – for you. And in order to get that you need a bit of virtualization technology in order to sub-divide or cut customers up in a different manner so that you can bundle solutions for that individual. I may not be a Facebook user but I may want ESPN and some other stuff. And the telecom is a perfect channel, perfect conduit to deliver me those services. Another good example of, which I think is probably less than 15 years away – maybe five – is personalized TV.

I mean, I'm ashamed of the industry of what they've done with IPTV today. It's a joke. It's just linear TV with a bunch of bundles on the back end. It's no different than what I can get from Comcast today. There's just no point. The whole advantage that telecom have over a cable telecom, for example, is an ability to offer individualized, personalized TV channels. I don't know why we're not there today. If I could get an EPG, an electronic program guide, from an telecom that pulled everything together: I would buy it. And I'd probably spend a lot of money. So I want all my regular over-the-air broadcast TV channels. I maybe want TV5 from France. I maybe want some Vodacom TV show in South Africa. I may want Hulu shows, Netflix movies. I want to pick and choose what I want just like when I boot up a windows machine and it gives you your profile. My desktop is completely different than my wife's at home. I got my own set of applications. My own Excel spreadsheets and Word documents. Why can't I have that exact experience on a TV screen? I don't know why. But the telecoms could be doing that. I don't know why they're not doing it.

THORSTEN: Isn't it interesting that cable companies have all those relationships with the programmers?

BRIAN: Yeah.

THORSTEN: And they could actually offer a la carte programming.

BRIAN: Yeah.

THORSTEN: But you can only buy them in blocks for obvious reasons.

BRIAN: But that's because the programmers won't sell it that way. They make you take it as a bundle. So if you go to Disney, Disney says, *"Oh, okay. If you want ESPN that means you got to have this card."* And the card has a bunch of channels on it.

THORSTEN: Right. But isn't that the same problem then for the telecoms?

BRIAN: Yes. I do think it's a structural problem for the industry. I think, today, the content guys have to change. But as a telecom, even if I'm going to buy the exact same thing as a cable company: I also should have Hulu and YouTube and Grandma's pictures and all this stuff all integrated in one UI. Why don't I have that today? It's totally possible.

THORSTEN: Because the commercialization aspects are obviously missing. In the end the question is always who's going to pay for it. You're talking about the air interface build-out and other technologies – who is going to pay for it? Or how are we going to pay for it?

BRIAN: We have to get to a point where the bundle pricing and all you can eat pricing disappears. You need to do tier pricing, per application pricing. So you link the application to a resource. So if I want to watch Hulu video on my cell phone, it's going to cost me a lot more than if I was going to do SMS. But we're not at that point today. And I think the FCC is actually a little bit too in favor of the web companies and not enough in favor or being able to differentiate on pricing, network infrastructure, et cetera. And what it'll do is it'll drive broadband infrastructure to a halt.

THORSTEN: So what you're basically saying is right now people are either completely for open networks, which might kindle innovation ...

BRIAN: ... innovation at the application layer. You get no new bandwidth ...

THORSTEN: ... and the other option you have is to establish cost measurements. But that doesn't give you any more revenue and is not going to pay for the huge investments you have to make on the air interface.

BRIAN: The only other option is for the government to basically take over the infrastructure. That's where it's headed. Why did they ever privatize AT&T? Why did they split it up? Because if they want open Internet and they expect people to pay for it – nobody's going to pay for it if there is no extra value. So the government is going to have to basically sponsor a bandwidth build out, which is what's happening in the world. I mean, if you look at the rural program, yeah, that's what it is. They're siphoning from the metros in order to give to rural.

THORSTEN: But the USF (Universal Service Fund) for rural build-out is getting smaller every year. Is this a dilemma?

BRIAN: I think it's a political problem. I don't think there's an easy solution. I don't think people have the will to change it at the moment.

THORSTEN: So the question I asked before about the 15 years: Many telecoms are now looking at fiber rollout or other network investments, like LTE or WiMax. A lot of these business cases are looking at a 15 year timeframe. It's very difficult for telecoms to understand the impact of current developments. Because it may be that in the minute they rolled out fiber the regulator is going to say *"great that it's there, now you have to open this for everyone."*

BRIAN: Yeah. At a wholesale price. At cost.

THORSTEN: Maybe only pennies of what they paid. If your portfolio companies decouple the fast moving application innovation from slow moving infrastructure build-out, infrastructure build-outs are still going to be a 15 year investment. How do you pitch your investments to telecoms, with an uncertain future and uncertain business cases?

BRIAN: This is a grossly aggregated comment, but the technologies I'm seeing being adopted inside telecoms today are very network optimization centric. They are not CAPEX related. If it's a technology that enables me to squeeze more, utilize more, I see those getting adopted much faster: How can I squeeze more revenue out of it? How can I squeeze more cost out of it?

On the pure fees and speed, Huawei is really making an impact around the rest of the world. It's made a huge impact, just because it's just a lot cheaper. So people are looking at it. I have X dollars per megabyte. And next year I'm not getting any more revenue but my cost better go down. So the only way to do that is by less expensive equipment.

THORSTEN: Interestingly enough, Huawei, in a lot of area – and Detecon conduct a lot of RFP's so we see what the pricing is –, is not the least expensive vendor. But if you look at the total cost of ownership, then they are the best because they are future proof. They are more willing to customize a solution. You can innovate faster. You can just exchange hardware more readily. They have a common plane on all the routers. There are a lot of things that are very attractive about Huawei.

BRIAN: To me the sheet metal business, from a venture capital perspective, I would not invest in a sheet metal and device company in the U.S. anymore. That business model is gone. I don't think it exists anymore. So the stuff we invest in is very software centric or, if it's hardware related, it's usually in Asia somewhere.

THORSTEN: The U.S. "sheet metal" companies seem to know that, too – everyone is going software. And we're back at the Cloud discussion. Thanks for sharing your time and insights with us.

BRIAN: My pleasure.

web
kind
sun
data
system
first
without
enterprise
applications
model
anyone
agencies
dynamically
using
infrastructure-as-a-service
management
amazon
requirements
meet
network
scale
benefit
telecoms
servers
world
industry
open
practices
computing
always
site
get
lot
way
able
see
things
one
application
much
networks
design
every
already
change
basic
platform
build
developer
resources
advantages
running
service
need
per
use
just
perspective
security
think
offer
getting
standard
cios
today
suddenly
cloud
building
provide
provisioning
others
provision
machines
many
developing
like
storage
want
people
advantage
now
level
clouds
customers
start
major
right
systems
important
public
seen
time
take
customer
technology
processes
run
something
traditional
quickly
operational
components
business
great
architecture
centers
best
built
costs
services
standards
center
different
demand
help
obviously
expertise
new
real
developers
companies
reason
also
infrastructure
looking
needed
manage
go
private
api
market
providers
virtual
interesting
really
going
telecom
government
made
therefore
capital
large
know
cost
actually
company
provider
make
may
well

Sun Microsystems
December 16th, 2009

Sun Microsystems is without doubt one of the largest telecom vendors world-wide. Sun's network computing infrastructure solutions are used in a wide range of industries including technical/scientific, business, engineering, telecommunications, financial services, manufacturing, retail, government, life sciences, media and entertainment, transportation, energy/utilities and healthcare.

In December 2009 Detecon's Thorsten Claus spoke with Lew Tucker, CTO, Cloud Computing at Sun Microsystems. Lew is one of the great visionaries of Cloud Computing, with hands-on experience how to get these visions into the market: He spent much of the '90s at Sun and contributed to the explosive growth of Java and growing Sun's presence on the Internet. In 2002, Lew joined Salesforce.com and led the design and implementation of App Exchange, which remains one of the largest cloud computing success stories to date.

THORSTEN: A lot of our customers immediately think "Amazon" when I say "Cloud".

LEW: Yes, Amazon has jumped into a leadership position in cloud computing by leveraging their experience running a massive infrastructure for their retail site, but when you think about the components necessary to run Clouds it might just has easily been telecoms to first offer Cloud services. They have the scale to benefit from the multi-tenancy characteristics of Cloud, the network, and the relationship with customers. While it's interesting to see Amazon as the market leader, I would have expected it to be the hosting service providers or the telecommunications and network providers.

THORSTEN: But isn't it the case that telecoms have the cash cows and they're riding them and because they're market leaders in their space? They basically wait until Cloud reaches a critical mass so that they have to move.

LEW: Cloud computing is all about networking. Telecoms can provide many network services into the enterprise along with the service provisioning that they do right now for their customers. By adding compute and storage servers, they can quickly make an impact in Cloud Computing

I think we saw Cloud Computing start at two levels. At the top level we've seen it kind of Application-as-a-Service based, like Salesforce.com for example, where I actually worked for a short period of time. That's where we've seen people getting applications over the Internet.

And then, most recently, we've seen it now actually go down to the most basic level – the Infrastructure-as-a-Service – where customers now are essentially able to rent computers online at very reasonable prices, in the order of ten cents per CPU hour; Dynamically provisioning, just those computers that they need.

THORSTEN: Right, but what is Sun's vision or perspective on Cloud?

LEW: Sun, all along, has been a driver in the evolution of the Internet. We envision a world of many Clouds, public, private and hybrid that are open and interoperable.

Sun plays in all these areas. We have a very large developer community, large enterprise customers and we work directly with a lot of large web providers. Cloud spans the range of technologies that Sun provides our customers – data centers, servers and networking. Cloud computing is a natural fit with Sun's business models.

THORSTEN: If you have such a broad array of customers what is the common fabric that Sun provides? I mean how can you cater to so many different industries?

LEW: At its core, Sun Microsystems is a systems company. We provide IT departments everything from servers, storage and software to technology and professional services, helping our customers build Clouds to meet their specific requirements.

Using Sun's Open Cloud Platform, along with our systems architecture, we are able to help customers and partners build and manage public and private clouds.

Many people today think about Cloud computing at the lowest level: getting on with a credit card and then you spin up your own server. Well, that may be what a developer wants to do. But a company, in fact, wants to do much more than that. They need to have their IT organization run their servers in the Cloud. So they really need groups of people. Then you get into much more traditional IT management systems and things like that that are running in the Cloud.

THORSTEN: You just mentioned tools to manage Cloud from an IT department perspective. That's something that I've rarely heard so far. Everyone is talking about Cloud and providing things, but no one talks about large scale management, about large groups.

LEW: Our belief that is that most customers want more than just a few servers on the Internet. They would like to have a data center in the Cloud which they can dynamically provision with as many or as few servers as needed. In addition, even in a small web startup, several people are responsible for developing and managing an application so we need to see advancement in role-based permissions and tools for teams to manage cloud resources much as they would in their own data center. Therefore, our model has been to center the design of cloud services around this concept of a "virtual data center."

In a virtual data center, all customer resources are virtual while the service provider supports and manages the actual physical infrastructure. In this way, customers avoid all of the headaches associated with maintaining that infrastructure, replacing the components, etc. They simply provision virtual machines through the web as needed.

THORSTEN: How would a telecom start? Or with what would a telecom start? What's my first step I have to take?

LEW: The first step is probably to contact their Sun sales rep. *[Laughter]*

Following that, telecoms should evaluate their needs and the needs of their customers (who may already be using the telecom's network or hosting service) and then focus on finding the right system architecture and design expertise.

Sun can provide the technology, architectural expertise, and even core applications such as email, collaboration, and identity management, and then customize each solution for our customers.

One of those customers today is AT&T, which recently announced its Synaptic Compute as a Service offering based on Sun technology. They are using our reference architecture, hardware, storage and Sun Cloud API.

THORSTEN: What does the [Sun Cloud] API do?

LEW: The Sun Cloud API is a basic API for self-service resource provisioning that we published under the Creative Commons license so that anyone, anywhere could use it. It provides a very easy way for developers to realize the benefit of a virtual data center.

The Sun Cloud API starts with the concept of giving each user their own virtual data center. Using this API, developers can provision servers, attach storage, assign IP addresses, and configure networks.

THORSTEN: All the basic functions.

LEW: Exactly. What's interesting is that there does seem to be pretty widespread agreement on the basic components that are needed at this "Infrastructure-as-a-Service" layer. Sun has been working with a number of standards bodies to discuss the viability of a common API and it's one of the reasons why we put the Sun Cloud APIs into the public domain. Anyone who wants to implement it or extend it in any way is free to do that.

THORSTEN: A lot of our telecom customers say exactly that: "I will start with a Cloud, but which one, and then I get locked in, I can't move out." Everyone is looking for a standard or standard body that basically creates a common, exchangeable ecosystem.

But from a vendor's perspective isn't that dangerous? Aren't you risking getting replaced once you are exchangeable?

LEW: Sun has always recognized the importance of open standards for our customers and our business. We believe that an open strategy leads to more market opportunity and makes it easier for customers to talk with us without fear of vendor lock in. In a growing market, all benefit.

Vendors compete on the implementations, the excellence of system architecture, and the other services. I believe this leaves plenty of opportunity for the various players to compete and differentiate their offerings. Agreeing upon a standard API will simply help the market expand more rapidly.

THORSTEN: And then the market is going to be larger, meaning your share is also going to be larger.

LEW: Exactly right.

THORSTEN: Many telecoms are coming from a walled-garden approach. They want to be as locked-down as possible, but they also want to have six different vendors. But that suddenly doesn't work anymore, especially now, where we see private Clouds and public Clouds merging, Cloud being expandable. Do you think that this is going to be the future …

LEW: Absolutely.

THORSTEN: … or are there only private Clouds in the long run?

LEW: There will be both some very large "public" clouds, companies building their own on-premise (private) clouds, as well as some hybrids. We've already seen a lot of companies that have moved entire applications such as CRM out to SaaS companies in the Cloud. Others would like to be able to tap into the resources available in the Cloud by extending their on-premise network to a virtual data center at a cloud provider. This would give the company more flexibility during peak demand and allow the company to avoid having to carry excess capacity on site. This has come to be called "cloud bursting" and the idea is that the customer could push some applications to a public cloud during peak times. The issue is: How do you do this safely?

How do you extend your network out into the Cloud? This is where network providers have a particular advantage. Whether it's MPLS or other kinds of secure means of extending a network from a private Cloud out into a public area, those in the telecom industry have the technology in place to make this happen.

THORSTEN: With all these opportunities at hand, what kind of stumbling blocks do you see in the telecom industry?

LEW: Right now, the major web companies Amazon, Google, and Microsoft are leading the charge. This is because they learned for themselves how to build web-scale systems and had to innovate quickly. The biggest question for the telecom industry is how quickly it can catch up, and how it can foster the kind of innovation that will be required. We are at the beginning of this era of Cloud Computing and much remains to be done.

Telecoms are in an ideal position because they own the networks, but will need to innovate to be major players in this market. They have inherent advantages in delivering network-based services such as private networks, load balancers, firewall services, content distribution networks, etc, but will need to move fast. Telecoms should look and see what services they are already providing and consider using the Cloud as another avenue to bring these services to the marketplace.

THORSTEN: That's an interesting view, because I think most of the telecoms are looking at one level higher. They say: "we have so many industries and verticals – we have banks, we have oil companies, we have manufacturing – how do we provide a platform for all industries? Telecoms are often organized in industry lines, and now every industry line is thinking about which platform they I offer on or in the Cloud.

LEW: There absolutely will be vertical, industry-specific plays, but if you start with a basic Infrastructure-as-a-Service platform, you aggregate all of your customer demand and can pass the cost advantages you get in scale down to your customers. This is what makes cloud computing different from the older hosting model which unfortunately inherited all the complexities of traditional data centers, and therefore all of the cost.

THORSTEN: So, where is that all going? Let's say we meet again in ten years or way in the future …

LEW: One of the more interesting announcements of late has been the US government's expressed interest in cloud. Vivek Kundra, the federal CIO, has said that many government agencies will use cloud computing.

To facilitate this shift, the US government is taking a cue from Salesforce.com's AppExchange site and building an apps.gov site, where government agencies will choose the Cloud services they need. Instead of building or expanding their own data center, government agencies will go to the marketplace and get Cloud computing services to run their agencies. With this initiative, I expect to see a government Cloud forming, which will hold the set of services, data centers and providers that can meet the requirements established by the government.

From a global perspective, even the World Economic Forum has expressed interest in Cloud computing and the hope it brings to help the developing world take advantage of the information and services to build their economies. It will be interesting to see if much of the developing world is able to take advantage of cloud computing and entirely by-pass the buildout of private data centers.

THORSTEN: Sun is one of the leading providers in the Cloud space especially with your activities, the openness, and the engagement on API standards, which is likely to push the whole market forward. How do you as Sun ensure security? How do you tell your customer, "You can really trust us."? What do you do or what are best practices telecoms should do?

LEW: There's a technical and a business processes aspect to security. One of the first hires we made in our Cloud Computing group was a Chief Data Governance officer – Michelle Dennedy. We recognized that security and data privacy issues were going to be paramount and we therefore needed to address these issues right from the start.

We also have been working with security organizations such as the Cloud Security Alliance that are developing best practices, upon which we could build technology, which incorporates these principles.

That's how we came up with the concept for Immutable Service Containers. In Cloud computing, a change has occurred in the way applications are often released to the web. Traditionally, application code, created by a developer, was handed off to an operations team which tested and subjected the application to scrutiny by security experts. The operations team was responsible for deploying the application and had guidelines for securing networks and servers to minimize security risks. Today, developers may simply pick up virtual machines made my others in the community and deploy their application without much thought. While this certainly speeds deployment, it is not without risk.

To make it safer for developers to deploy applications (without becoming security experts themselves), we've taken all the expertise and best practices from the security realm, and built it into virtual machines, upon which the application code can then be deployed. This is how Immutable Service Container virtual machines were born. As a design pattern, companies can make their own as well, building in their own policies and best practices.

Immutable Service Containers for OpenSolaris are now available for anyone to use on Amazon's EC2 service as are other open source security tools that we've released to the public.

THORSTEN: We saw a couple of companies that had a private Cloud, and it failed. The reason why it failed was because their internal processes for security testing, approval, and setup were very complex: They were concerned and didn't know what's actually going to happen to their internal infrastructure if this service would get a huge internal uptake and what that would do on their production system with actual customers on it. Because the processes were starting to get so complicated, no one was using the Cloud. All the benefits you had with Cloud – poof. Away.

LEW: That is a real threat. Right now, regulatory requirements, existing IT policy, and government requirements were built for traditional data centers that were behind firewalls. They have to be adapted.

Forward looking CIOs are recognizing that just as PCs came into the enterprise through a line of business and individual users, Cloud Computing is having an impact in the same way. The sooner CIOs embrace Cloud and recognize the need to look at governance and best practices, and engage the line-of-business people who are using the Cloud the better. CIOs need to become a real partner and trusted advisor for departments who are ready to engage in this new style of computing.

In fact, I've been very pleasantly surprised by how quickly large enterprises have become interested in Cloud. I would have expected larger companies to lag behind as they traditionally have with new technologies.

What the enterprise sees is real cost advantages, particularly when the market is now pricing computing at less than 10 cents per CPU hour, 15 cents per gigabyte per month stored. CIOs are now looking at their own data centers and thinking, "How is the market able to provide these things at a fraction of the cost that my own department charges for these services?"

The reason is that Cloud Computing strips away a lot of the unnecessary complexity that has built up over time and moves towards a more centralized, standardized model for physical resource management. Then it uses virtualization and self-service to give end-user customers the flexibility they need. This is a win-win for both the IT service provider and the end-user.

THORSTEN: It seems then that the architecture and design and processes are much more important that the actual IT.

LEW: I've always been a believer that architecture matters. You can have great processes and a weak architecture, and it'll leave you vulnerable to security and reliability issues, among others. It always comes down to architecture. One of the advantages of Cloud computing is that it allows you to experiment more quickly with different architectures, because you can quickly bring up 50 virtual servers to explore how you might partition a database in a particular way, run it for a couple hours, and do some tests. When people ask how do I get started, I keep saying, "Just try it." The cost of doing your own experiments is so low there just isn't much reason not to.

THORSTEN: I have an interesting customer right now with one of my startups that I'm funding where we started on a public Cloud, because that's the quickest way to start.

And if it would still continue on that Cloud, we would sign a check of about $90,000.00 every month. And we have a lousy 500,000 users, not like the 300 million on Facebook, right? That cost is quite substantial. A 16 terabyte Network Attached Storage that can do the exact same thing costs me $5,000.00, the collocation another $1,000.00 per month.

So, you compare $6,000.00 with $90,000.00 and you think, "Oh my God, of course I go with the collocation and NAS!". However: who is going to maintain that? Who's going to scale that? Who's going to build it out? What if I need two more racks in there and that the collocation is just not capable of doing that? Suddenly, I need to pick up a GBit WAN service. I have a software upgrade to distribute across all servers. I need to harden all the services …

LEW: If you're just looking at the equipment costs, if I know precisely the demand, and it never changes, it's almost always cheaper for me to buy it myself. But we're forgetting the operational costs, and the opportunity costs that you give up because you are spending all this time maintaining your system.

THORSTEN: Amazon has one-price-fits-all. But obviously, if you go to GoGrid or anyone else they can provide customized [Service Level Agreements], customized customer service …

LEW: Yes, other companies are rapidly moving into providing Cloud services, which will create a lot more competition. Rackspace and others, emphasize great customer service.

THORSTEN: One of the problems telecoms have is that traditionally business units had a [Capital Expenditure] budget and then operation was done by a different [Profit and Loss] center. That operational expense, in many cases they never saw that. Now with Everything-as-a-Service you suddenly have a very low [Capital Expenditure] and a very high [Operational Expenditure]. The whole dynamics and business mechanics internally are changing.

LEW: When businesses start to get most of their computing needs through the cloud, the way in which they treat capital versus operational expenses will undoubtedly change.

Seasonal businesses that previously had to over-provision for peak demand will obviously benefit from this new model. If I'm building up something for the Olympics – that's only every four years -- I no longer need to carry that capital expense when I'm not using it.

THORSTEN: Is there something missing where you say, "Gosh, I wish we would have that!"? And who would be a potential provider of that – besides Sun, obviously?

LEW: I think that we already touched on a number of these things. An agreed-upon standard for Infrastructure-as-a-Service is the big missing element today. Secondly, we are recognizing that security is a major issue and that storage of personal information has associated privacy regulations that may differ for each country. Knowing where storage is physically located therefore becomes important. This may require us to always associate location metadata with each stored record. On the enterprise side, companies are looking for better *Service Level Agreements* and I expect to see competition between the different service providers around what kind of SLAs they will offer.

THORSTEN: Very interesting. What's one of the exciting applications or technologies you see – one that you can talk about?

LEW: One area where I expect to see some radical changes is how the use of dynamically created virtual machines will cause us to change the way we design applications. With this new model, we will start to built applications as systems that are in some sense, aware of their own resource utilization, and can take steps to either increase the number of servers in response to increased demand, or decrease their resources to save money.

Instead of having to file a trouble ticket with a system telecom, saying "Please spin up four more servers for me," now the application itself can do that provisioning through APIs. Of course, guardrails need to be put into place both internally and with the service provider because you don't want that application to go and run up a $50 million bill in a week.

THORSTEN: And building business rules …

LEW: Yes, one can build in both business and economic rules into the application itself.

In fact, the application might also be able to notice, "Gee, I'm very lightly loaded. I'm going to save some money by turning off some of my resources that seem to be very lightly loaded."

THORSTEN: Interesting. Do you think that is going to be managed by the application itself, or rather by an application overlay or application controller – some sort of a business policy server?

LEW: This approach does require some real careful consideration. We all know the Skynet scenario in the Terminator movies which shows what can happen. But we are approaching this capability, so now is the time to start thinking about how we make applications aware of their true cost of running and build in economic rules and the necessary oversight.

Service providers will also need to look much more at monitoring application behavior before provisioning resources. If an application all of a sudden is spinning up 1,000 servers that may want to be blocked and a call made to the owner to ask, "Is this a mistake? Or is this something that you really intend to be doing?"

THORSTEN: So real-time customer care suddenly becomes really important.

LEW: Absolutely.

THORSTEN: Of course customer care has always been important. But the real-time aspect clearly shows the intersection of automation, self-service portals, and real-person interaction.

LEW: Again, I think this is something telecoms actually understand. Support and service are going to be some of the real differentiators.

THORSTEN: Let's say I'm a large chemical company. I have a carbon footprint. I'm running applications. Suddenly applications spin up servers themselves. Do they need to report and monitor the carbon footprints as part of the running costs as well?

LEW: Could be. What we know is: the main issue for computing in terms of the environment is that most of the servers running are so under-utilized, and therefore wasting a lot of energy. The extent to which we can increase utilization and turn off servers that aren't being used in the data center, is how we will really lower the overall carbon footprint. I do believe that service providers will do a better job of doing that and managing that than your traditional data centers who may not have the manpower or the expertise to do it.

THORSTEN: Great, well thank you for the time and insight. It's been very interesting.

LEW: My pleasure.

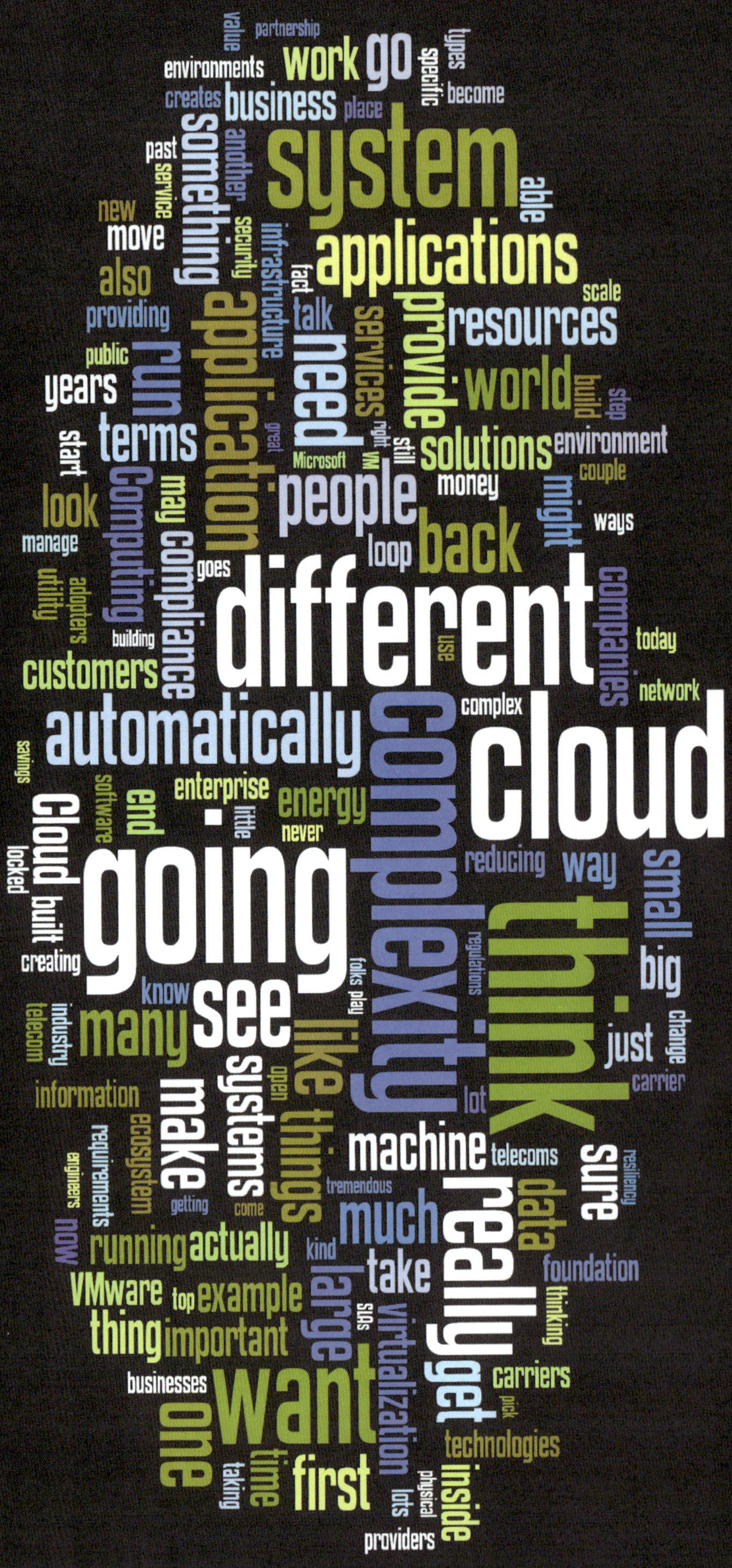
partnership
value
types
work
go
environments
specific
become
creates
business
place
system
past
something
another
service
able
new
security
infrastructure
move
applications
also
fact
scale
providing
talk
application
provide
resources
public
services
run
need
world
build
years
step
great
right
still
terms
environment
VM
Microsoft
solutions
start
couple
may
people
money
look
Computing
might
manage
ways
compliance
loop
back
utility
adopters
goes
different
companies
building
use
today
customers
complex
network
automatically
cloud
complexity
savings
software
enterprise
energy
end
little
Cloud
never
locked
reducing
way
small
going
built
think
regulations
creating
folks
play
big
know
see
change
telecom
industry
many
like
just
open
carrier
information
systems
make
things
lot
ecosystem
requirements
machine
telecoms
sure
resiliency
engineers
tremendous
data
getting
come
really
much
now
running
actually
kind
take
foundation
VMware
top
example
large
SLAs
virtualization
thinking
thing
important
get
carriers
businesses
want
pick
one
technologies
taking
time
first
lots
physical
inside
providers

VMWare
April 9th, 2010

The Fusion Red roadster came out of nowhere and silently whirred past me out of sight. Driving to VMWare's Worldwide Headquarters in the gently rolling hills of Palo Alto next to Page Mill Road, I was once again reminded that Tesla Motors is just around the corner, but you will also pass HP Labs, FX Palo Alto, PARC, SAP Labs, and a few others of the most innovative companies of the Valley. VMWare fits right in. Its virtualization solutions won countless awards, helping businesses to steer clear from adding new applications on incompatible architectures, across disparate collections of hardware and infrastructure.

In April 2010 Detecon's Daniel Kellmereit spoke with Richard McAniff, VMWare's Executive Vice President and Chief Development Officer, about the difficult tasks of telecoms in getting Cloud solutions up to speed in security, reliability, portability, and in matching shared resources to Service Level Agreements (SLAs).

DANIEL: Thanks for taking the time to talk with me today. Can you start by explaining a little about VMware's role in Cloud Computing and the Cloud Computing ecosystem?

RICHARD: To answer that question, you have to think back to where we come from and what we are all about. The fundamental foundation of what we do is virtualization. That is really what we have built the company around, and what has transformed an industry. And when I look at that I say: Wow, that is an amazing thing that VMware has done. They figured out how to do that – that was actually before I got here.

That still forms the foundation of where we think computing is going – that foundation of virtualization. How do you use that foundation to really improve overall productivity and reduce complexity? Cloud Computing is not so much a destination per se, it is really much more a set of principles by which we think through how we develop products. It is about reducing complexity and how you provide elasticity, pay-as-you-go and make things more automatic so people don't have to think 'how do I sign up to get more machines?' We are so used to going to the Internet, for example, and there are an awful lot of things we can do.

So, what are some of the characteristics that we are thinking about? First, how do you take this foundation of virtualization and make it much simpler so that it automates itself? Second, how do you take information and automatically channel that information to a feedback loop that actually automates the system for you. So, that is where it starts and then we build on top of that.

DANIEL: Let's step back for a moment to the big picture. How will Cloud Computing evolve in the next couple of years, and what are things we might be talking about in 10 years?

RICHARD: So let's remove this from VMware for a moment, so that this is not a commercial pitch about VMware. I think that we can talk about what VMware is doing, but I think it is as important to really think about where the industry is going and what we are actually trying to participate in.

I think there will be a world in 3-5 years that is probably a high grid world, where many people continue to built datacenters that take on characteristics that you see in the public cloud – elasticity, pay-as-you go, on-demand, open resources – things that you would think about if you are thinking about a Cloud. You'll see that inside the enterprise and you'll see that outside the enterprise as well. And I think you will see the ability to bridge between these different environments, where some people will build applications that run inside the enterprise, inside the private cloud behind the firewall. But there is also going to be something in the public cloud that you can reach out to and also run applications on.

You'll see different companies addressing that in different ways. Some of the companies will be more aggressive and some won't, because there is lots of compliance and security where the data has to be and all kinds of regulations and rules and so on that you have to take into consideration. And that will drive different companies in different ways, based on those particular regulations and how they affect them. So, you are going to see a world where you will have inside the firewall cloud-like datacenters and you will see outside the firewall cloud-like infrastructures. And then you are going to see a bridge between those two different environments. This is what I believe you are going to see in 3 to 5 years.

DANIEL: So how will businesses change through further introduction of Cloud technologies? Will Cloud just become a utility that everyone uses like, let's say, electricity in the past? What will the transformation look like?

RICHARD: Well, at the logical extension of where that goes it will be like a central utility that you plug into. There was a time when everybody had their own power plant. If you wanted to be in the garment industry, you first had to build a power plant before you actually build your factory. And then over time we discovered this thing called AC current. We were able to go into different types of utilization and to a central model. Part of that happened, I think, because of natural constraints. You are going to have people inside the datacenter and also internally in private clouds and so you will see those exist for a very, very long time.

I think, getting back to the idea of Cloud being a utility, in some senses it may look like a utility, but in some cases you may run it: you own the CAPEX and you own the OPEX and you run it. But it is not going to feel that way. Because what you are going to do is, you are going to go to a catalogue where you just say: Hey, I need additional resources! Or, I run an application and that application automatically can scale, because it automatically gets the resources it needs. You don't have to think about it – it becomes an automatic thing. So again, over time you may even see people get rid of these terms 'private' and 'public', which will become 'The Cloud'. Because the characteristics are really the important, rather than: Is an outsider involved.

DANIEL: Let's talk about adoption of Cloud technologies. You mentioned that big enterprises are looking at it. I had the chance to interview Aneesh Chopra, the Federal CTO; he is very bullish about Cloud. He believes it might be one of the next big growth engines and he believes it will have a similar impact on the tech world as it will on other areas of business. So, who is going to be the first adopter? There are all these small and medium sized companies that do not have the legacy and maybe as many compliance issues as you mentioned, there are large corporations, and there is the government as maybe the most complex of these groups. Are small companies going to be the first adopters, followed by larger companies, followed by the government? Is there any kind of "rule of adoption" you can observe?

RICHARD: I think it is a little bit of everything. You have to look at how people are adopting Cloud today. Often, we forget that a lot of the applications we run internally that make up organizations and enterprises are already Cloud. They're pulling out their CRM or HR system or financial systems and so on, and moving them to the Cloud. So are they the first adopters? They may not think that they are adopters but they are already using Cloud applications.

You see small businesses that are drawn in towards the Cloud because they do not want the complexity and they do not have their own IT organization. So you can see those folks move as well. You are seeing a lot of early adopters in the enterprise in terms of building out applications and then pushing those applications out. So, there are all those different places and of course the government and the public sector is also going to move there too, because of the efficiencies and the agility and the reduction of complexity.

That is why I said in the very beginning, we go back to our roots of virtualization. If there is a common thread that we think about in terms of how can deliver that value, it is reducing complexity. At the end of the day, how can we reduce complexity? Everyone is trying to do the same thing in terms of how do you become more efficient, reduce complexity, drive the cost out of CAPEX, drive the cost out of OPEX, and become much more agile.

DANIEL: On the application layer, is there something we can do with Cloud that we couldn't do before? Are there specific application scenarios that are unique to the Cloud, or is there something possible now that was not possible, let's say 5 to 10 years ago?

RICHARD: I think there are a couple of things. One is that traditionally, to build out an application, it can take a very long time. It can require many, many, many signatures and a lot to roll this application out. You need to get the machines, you need to get the signatures, you need to get the application deployed. In this new world in many ways it is on-demand. If the application is deployed, it automatically scales, it is managed, and as a user, I pick that application and it starts to work. And on another level, it is the same thing: If you want a machine, you just say 'I need this machine' and then the machine is automatically provisioned for you. But it is a virtual machine, it's not a real machine, and it's automatically provisioned. So, that is a big change, again, and people's thinking will change.

Now, there is a whole different angle to the question, which is: How are applications being built? They are being built differently from the way applications were being built in the past. But that is more of a nuance of how data systems work and underlying services work.

DANIEL: George Bernard Shaw made the great statement that science never solves a problem without creating ten more. So what are we all running into here? Everything becomes easier, faster, more agile, but where are some issues and complexities Cloud is creating that need to be solved?

RICHARD: The key to all of this is making sure the data is secure and it is compliant and that we do not get ahead of ourselves in terms of saying 'It's a really great new world' and we forget about the fact that privacy is still paramount. We have to solve some of these problems that are different than what we had if you have everything centralized in one place, when you knew where and how to manage it. I wouldn't say it creates new problems – it really creates new challenges and opportunities that go along with those challenges.

DANIEL: In complex systems, complexity often stays constant. If you have complexity on the one side and you want to take it off, it creates complexity in another area. How does this relate to complexity in the Cloud? Are we taking complexity away on the user frontend, but creating a lot of complexity in the back end?

RICHARD: No, I wouldn't exactly see it through that lens. What I would say is: If you have a device like the iPad which is a very easy to use device, inside it is still very complex. It's a marvel of engineering technology, yet it still can be managed and easily used. So, while we made something that is much smaller and much lighter weight with touch screen and so on, we did not actually add tremendous complexity to the system, we took complexity out of the system.

Let me give you an example of what I mean in terms of virtualization. Think about scheduling and where you place VMs (Virtual Machines) and how you move VMs around. You are actually removing complexity from the system, where I do not have to worry as an administrator: 'Oh, I am running out of resources over here. What do I do?' A VM automatically finds those resources you need, and automatically assigns those resources, in order to meet SLA or quality of service requirements that you might have. So it is removing complexity from the system, but it is not introducing any other complexity. Now, it does introduce complexity to my engineers and my engineers have to be amazingly, phenomenally great engineers because this stuff is really, really rocket science. This is really hard stuff – HA – which is what we call 'High Availability'. If something drops, we just re-create it via VM. In a physical world, that physical machine would go down. You have to send a ticket to your helpdesk, someone has to pick up that ticket and find that machine and reboot that machine. We do all of that automatically, and that is an example of taking complexity out of the system and what that does. Of course, it removes the OPEX and drives savings for the organization. So it is energy savings and how much energy savings you see when you start to pack lots of VM into one physical box. It is a tremendous opportunity along those lines. We are not creating other dis-locations somewhere else.

DANIEL: I agree, energy saving is increasingly an area of interest for large service providers and telecom operators. Many telecom operators are one of the largest energy consumers in their country. And most of the energy is spent on running the network, the core and aggregation network, and the access network. The further you move to the end consumer and CPE's, the higher the energy consumption, especially with new technologies like VDSL and GPON.

I want to speak a little bit about reliability as this is another area that operators are specifically interested in. Most solutions today are built with many parties involved; going forward it will more and more be kind of a big mash-up of different technologies and vendors. So who is going to manage the quality of service and the reliability of applications? This is something that many of our clients are thinking about, who is going to be the guy whom I can call up, who is managing my SLAs?

RICHARD: The key is understanding how the system performs and what is going on with the system. Am I running out of resources or are my resources adequate to match my SLAs?, SLAs might be different for different types of applications. In one case you may have a production application where you say I need seven 9s. This cannot go down – because now for 45 minutes to an hour I am going to lose millions of Euros in that process. So I have to make sure I have redundant systems, I have to make sure that I have back-up systems, I have to make sure that I can have disaster recovery systems that automatically will all work. And I need to know how the performance of the application works.

That is – again – looking at all this data through a different lens. Same thing with security. So it is incumbent upon folks like ourselves to take that information and provide that information to a dashboard or to the administrator and so on. But there is more to it. You have to be able to feed that data into a loop, where that feedback loop really says 'How do I actually tune the system'. If I am out of compliance, get yourself back in compliance. It just automatically happens, it shouldn't be something where I say 'Here is a ticket: We are out of compliance, now I got to go figure out what is going on.' It should say 'You are out of compliance. I have to make sure you have a report of that, so that we have a log where we track the fact that somehow, that something changed in your system, there was a signature that we saw, indicating there was a change'. Well, ultimately that comes back and feeds this loop that, in fact, automatically creates the system to be in compliance. Same thing with performance information that is going to be fed back into that loop.

We also have a large ecosystem of people who manage these systems and we believe that it is super-important for that ecosystem to work with our systems. Because we are not going to be all, we can't do it all. There are many different things that need to get done and it is very important for us to have an open system that people can use. There are certain things that only we can do because there is a loop, for example of software inside. And HA was again a simple example of how that works, where, when a machine goes down we automatically bring it back up. We are not losing anything. We should talk about our philosophy of being open and how important that is.

DANIEL: So let's talk about carriers and the role they should play in the Cloud. We have seen by now that many of our clients are really scared of becoming a dumb pipe. That is always the big first threat to them, and they have seen it happening in a couple of areas –mobile data is one of them – some of the high end phones now are going to be sold directly to consumers and in the end the carrier provides the SIM card and data contract, and maybe one or two apps. What if the same happens in Cloud? Google, Cisco, Microsoft, Amazon and others are ramping up Cloud infrastructure and resources worldwide. So will carriers in the end mostly be sort of an access pipe? They want to be more, provider of end-to-end solutions. But they are not moving at the same pace as some the innovation leaders, specifically here in the Bay Area.

RICHARD: Personally, I think that they can provide services on top of a platform. They do not need to be dumb pipes. They have specific value-adds that companies like Microsoft will never provide. I was talking to a large telecom recently and they were saying how much money they make on ring tones. So I do not know that that is the business Microsoft would like to have, or other solutions people. They have a relationship with their customers and they should continue to have that relationship with them. The partnership that we look for is that we can really provide tremendous value by reducing complexity. Providing resiliency and reducing energy consumption, reducing CAPEX. And I think these are all these things that we can do in a partnership with telecoms. On top of that the telecoms can plug in their value added services. We view this as a partnership, because we are in a little different game than some of the other cloud providers.

DANIEL: So when providing Cloud infrastructure and services to your clients like a carrier does, is it important to be one of the top players in terms of size? Are economies of scale very important? As a carrier, is it an advantage, if you can invest 5 billion dollars into Cloud infrastructure vs. a small amount of money? Do you have to be large and invest large amounts of money to be successful in this game?

RICHARD: I do not think you have to be large. There are different types of market places that will demand different types of service. Let's go out of the telecom arena and just talk about a small ASP that might be in some small city that provides value-added services to the school system. The school system feels very comfortable working with a local partner and they are building an application for example to manage their schools more effectively. In fact, I have talked to such a services provider recently at a partner conference. That is exactly what they are doing. They are aligning in a very interesting way with developers who are providing applications with VMware running the Cloud, accommodating small business in their city. So, there are economies of scale there. And of course the large telecoms are going to come out in a different way. If you are going to get into a certain business at scale, that is a different business. But there is room to play on many different levels.

I also think there are opportunities all over the world depending upon regional requirements. There is a very large ecosystem of smaller players that are like what you would think the old software folks were in the past – where they start building and delivering these solutions in a very specialized way to small businesses right around their environment – where the small businesses have specific requirements that generic solutions are not going to be able to handle.

DANIEL: Let's completely jump out of the highly developed world we live in, and speak about emerging markets. Some of the carriers in these markets have never even touched Cloud technologies. They run fixed and mobile access networks and provide a couple of value added services. In such a kind of clean slate situation, what would be your suggestion on how they should start engaging in Cloud Computing? What are the first things such a carrier could do if you can basically start from scratch, right now? Do you have any specific advice for such a client?

RICHARD: At a very fundamental level, I think they could offer Cloud Computing in their own operations. Instead of buying ten machines they can buy one. If you keep on going back to that first step on the journey towards Cloud Computing, you keep on getting back to this kind of virtualization. So without knowing what their requirements are, without wanting to suppose too much here, I would say just virtualizing their hardware as they deliver solutions to their customers is a big first step. It is saving them money instantly. The next step could be to define Cloud Computing. What does it mean for them? What are the applications that their customers are demanding? I know that we can save money on CAPEX/OPEX and I know that we can get increased resiliency. I know those things are the first steps for improvement that we look for when we look at an emerging market.

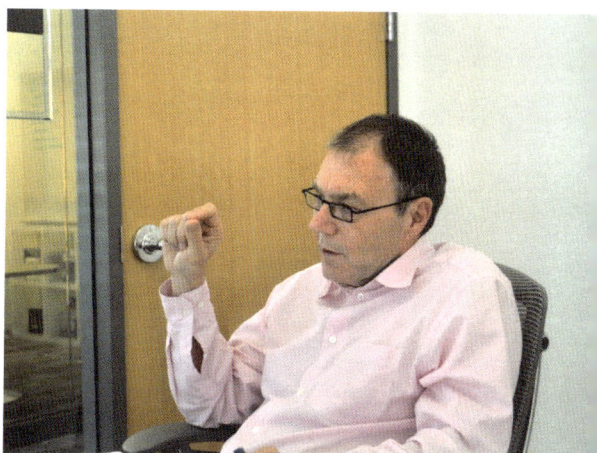

DANIEL: I would like to come back to the question of openness that you addressed in the beginning. That is something very relevant for large carriers that we work with. They do not want to be locked in to one system, one platform, one ecosystem, they prefer a very open approach with lots of room for navigation as this industry evolves. And they want to be able to monetize solutions their way, and not be locked in to a relationship, where they lose let's say 30% of their transactional revenues to a third party. What is VMware's position in this regard and what can we expect from VMware in the future?

RICHARD: Well, let's look at this first from the other side, because enterprises also want to make sure that they are not locked in. They want to be able to have choice as they pick the appropriate Cloud. Part of what we do to is to make sure that we are able to provide bridges to these different environments for the enterprise itself, so that if you want to run different kinds of Clouds we need to make sure that we are providing that ability. In terms of the solutions providers, I think it is important to provide them with some infrastructure that they use, but then you have to allow them to extend that infrastructure as they see necessary. And ultimately they can choose to be a development shop, they can leverage software micro-cells and at the end of the day, if you are using our software, they might want their own data bases. We are not selling databases. They can put their data where they want. So there are lots of ways that we can work with big telecoms so they are not locked in on core parts of their business.

DANIEL: That might be the right way to work with carriers, to be the enabler but not try to influence their business too much…

RICHARD: Exactly, we don't have to own them.

RICHARD: One other thing to think about that virtualization really does is that it isolates the operating system and the application. So you can choose whatever app you want to run. If you are providing Exchange, for example, you can do that. If you want to provide another email system, you can do that. If you want to go on a Linux system, you can do that. If you want to run SharePoint, you can do that. If you want to run Oracle, you can do that. If you want to run Sequel, you can do that. These are all choices that they have. In all cases OPEX/CAPEX goes down and resiliency goes up. We do not play in that (systems) space, we isolate them from that. And that is another place where they can get choice, where they can say 'OK, some of my customers are running this, some of my customers are running that and so on'. Whatever your customers needs, we are not driving you into a Linux environment, a Windows environment or any other environment. It is really up to your customers to determine what environments they need to have virtualized.

DANIEL: That is interesting. So you withdraw yourself from all the technology wars and opinions out there, the fights that take place between different providers and the partner network they are trying to defend.

RICHARD: If you really looked at what is our aspirational goal – it is really to be a trusted advisor. As this transition happens over the past several years, there are some people plugging into the wall today and getting the utility power, if you may. This is a transition that is going to go on for several years. Some people, as we said in the beginning, will never completely move because of regulations, security and so on. Some people will move more aggressively, depending on their market. It is a complex world to navigate in, in terms of figuring out the best economics, making sure you are in compliance, and provide appropriate security. And if we can help in being the Trusted Advisor to that environment, I think, then we have succeeded.

DANIEL: Thank you for taking the time and providing these valuable insights.

work
run Cloud better clouds
look sure home experience applications possible
Yahoo now resources division expensive
another great interesting already thing use everyone
devices mobile core Right
flickr like people DNS
sell general time potential using got
becomes end user go data
maybe telecoms Wi-Fi
issue network etc throw future base first
everything service going
need IP else get MapReduce
software phone actually different One
start know business way also
Hadoop idea many Internet infrastructure networks access
Amazon think cloud Mini information problems together
value years right things want companies
reason Opera Yahoo's customer
cost make stuff services interested beyond
kind good one Google back
device something away
might trying even web Coast open
channel example much problem five
difficult customers whether technology
code phones size probably cell clients
build server
wireless large create location
eleven connect routing

Yahoo!

May 5th, 2010

No, you don't really hear the Yahoo yodel in every cubicle and office at 701 1st Avenue in Sunnyvale. But serving almost 600 million unique visitors globally every month – about 48% of the online population – and 4.4 billion visits per month in the U.S. alone could make you yodel. While Cloud seems the obvious choice for efficient infrastructure and service management at that scale, Yahoo's decision to create, expose, support open Cloud services and technology enablers is less obvious.

In May 2010 Detecon's Thorsten Claus spoke with Tom Hughes-Croucher, Technology Evangelist at Yahoo, about how Cloud helps Yahoo to become the center of people's online lives, and how telecoms help service providers like Yahoo to deliver personally relevant, meaningful Internet experiences.

THORSTEN: What is your position at Yahoo? What do you do?

TOM: I am a Technology Evangelist.

THORSTEN: For which technologies? Any sort of technologies?

TOM: For quite a wide range. We do a lot of stuff with web services. We want people to embrace our web services, because our web services have a direct consequence on our bottom line. If you look at applications on cell phones or smart phones, almost all of them allow uploading photos to Flickr. And that is really great for us because we now have a business which is being supported by all these phones. If actually becomes a standard: If you don't do Flickr you are not doing photos on a mobile phone right.

For us this is really critical as it pushes large audiences to our site. One of my roles is how to get people to embrace using our services. Because if something like Flickr becomes integrated into peoples' lives, they go to our site more, view more of our advertising, and we make more money.

THORSTEN: So your revenue structure is focused on advertisement?

TOM: Flickr is kind of a strange example for Yahoo because it also has a pro subscription for more storage and API access, etc. But on something like Yahoo Mail it becomes more obvious that it is advertising. But there is another side, which is the open source stuff. That is more cloud related. This is one of the dichotomies that we have right now: We want people to use our services because they plug in to our audience business. What I mean by that is that Yahoo's unique selling point – with Flickr, Yahoo Mail, Yahoo Finance, etc. – to our advertisers is the fact that we have hundreds of millions of users. On the flip side, the cloud division is not interested in trying to sell people a service or trying to get people to adopt a service. What the cloud division wants is to share a technology they created in order to get more contribution inside of Yahoo. And the reason for that is: We are not selling our cloud software. We are not a software vendor. We simply want the best cloud software to build great web sites.

THORSTEN: But how do you then monetize your efforts within the cloud division if all you produce is APIs or tools, simply put?

Tom: If you read Yahoo news in a great app on your cell phone then you will also read Yahoo News when you are at home.

Thorsten: Got it. And how do you measure that?

Tom: A lot of people have a log-in experience. We have other tracking methods that center around consolidation. We don't talk about the monetization impacts of the APIs publicly. Internally there are some significant case studies that show that having a really successful API will multiply your business on your web site.

Thorsten: Flickr is again an interesting example because you are able to charge for API access. A lot of other APIs can't do that because it would prohibit developers to come to the table.

Tom: Because we can doesn't mean we do. All commercial usage of Flickr is a business relationship. In general, the reason why we implemented a charging mechanism was actually just to stop people taking value out without putting value back in. So in the majority of cases the relationships were synergistic. My understanding is that there isn't necessarily a financial arrangement. Where as in the case of Canon using Flickr for data mining it is like Canon is taking value out, but they are not creating new value.

But from a standpoint of the cloud division it is about 2 things: It is about saving cost and improving efficiency. It is not about 'can we create an EC2'. We want to commoditize what we make and share it as open source because that gives us a real ability to work with other companies that have similar problems. Hadoop is a great example. We have worked on Hadoop in the open and it has been incredibly successful. Lots of people are putting stuff on top of Hadoop. But we still contribute a large amount of the code base, say 90%, which is an awful lot. At the same that means there have been 10% of the Hadoop code base that we did not contribute to. 10% of bugs that were fixed by somebody else...

Thorsten: ... which might also be the 10% that are really important.

Tom: That's right – And that is the thing: It is an investment in the future. We can hire people that write Hadoop code already. You can't do that for MapReduce because it's basically proprietary infrastructure. You can't take it somewhere else other than Google.

From a long term perspective, if you go out four to five years, we hope that everyone uses Hadoop. Some people familiar with both Hadoop and MapReduce say that Hadoop is maybe a year or two behind MapReduce in terms of technology. Google had already been building a MapReduce infrastructure for two or three years before they released the service, so they were already ahead and poured a lot of resources into it. Yahoo has started pouring a lot of resources into Hadoop. And other companies have since joined us. Google has probably put the most resources into any implementation of the MapReduce concept. But at some point that will change. You work with a lot of telecoms, banking, finance industry, governments...

Thorsten: ...and suddenly the Fortune 500 companies invest into Hadoop. The game changes.

TOM: There is no way that a single company can compete with an industry or several industries.

THORSTEN: But Sun actually ran into a very similar problem. They were first with Java and first with a lot of things. They did a lot of open source stuff and then they were acquired by Oracle. Now you can argue whether this was a successful exit …

TOM: …but you need to look at their model, which was different: We are not trying to sell software. It is not our business. We sell services to consumers; we do not sell services to businesses. And this is one of the things that I am trying to do with cloud right now. I actually feel that we need to split the conversation. There are two conversations: The real reason that people are getting interested with clouds is because their IT departments cannot fulfill their needs: "I want to provision a server, I want it now." The best any IT department in the world can do is probably give you one tomorrow.

THORSTEN: Most of the time it is probably more like a month. And then you have to fill out this paperwork, stick to the processes and then, by the way, it is only going to be in this one location and you will have to physically be there, otherwise it is not working.

TOM: With cloud you have other sorts of benefits too. I want a server for a couple of hours to test this idea and then I do not want it anymore. Re-installing applications and creating permissions on traditional servers is really hard. But you crank up a virtual server and you can do all of that stuff. That stuff is interesting but everyone is seeing from the point of view of 'I need to go to somebody who already has all of this running and I need to rent it of them and it is a service'. If you talk about cloud, everything is a service. You think of Amazon, you think of Rackspace. There is a whole class of customers, big and small, that can afford and predict what they need, so they can and will build their own clouds. There are lots of good reasons for that. Yahoo is one of these organizations. We have the resources, skills, and the knowledge to build out our own cloud.

THORSTEN: Computing and computing resource management is one problem companies have. But in telecoms we see an increasing demand and complexity in service production. Within your infrastructure ecosystem you have a plethora of different end devices – 43 different mobile phones – and end device classes – TVs, BluRay players, gaming consoles, PCs, netbooks, cars, tablets, etc. All of these devices use different technologies and have different requirements. It might be a good idea to load 1,000 contacts in the background if you're sitting in front of a PC with a high-speed Internet access, but it's a bad idea on a mobile. HTML5 and its media objects as well as SIP messages might exceed a mobile networks' MTU[13] size – Ethernet has a fixed MTU of 1500 bytes, but some mobile carriers we work with set their MTU size to 1,300 byes. So you create a lot of chatter and noise. On the iPhone you also want to give back results as pLists – binary XML files – because

---

[13] The Maximum Transmission Unit (MTU) is the largest packet that a given network medium can carry. Ethernet, for example, has a fixed MTU of 1,500 bytes.

their parser for human-readable XML – serialized ASCII – is much slower. And user interfaces going to change with 'touch everywhere'. How is that affecting the interaction model? What do telecoms need to do in order to build and provide a service infrastructure for cloud beyond computing?

TOM: This is actually very interesting. Take a look at my website, http://www.cloud-peering.com/. All of the datacenters that everyone uses are roughly in the same place, because everyone builds datacenters in fairly known clusters. If everyone is in these places together, they should start to connect their clouds together and start doing explicit routing. "If you need to talk to some other cloud, you can do that and here is how you connect them together." For example, we want someone on Amazon to use our web services as fast as possible. How do you facilitate that? The way that we do that is to say: If we have these clusters, can we stick the clusters all together? Every one of these players has their own backbone. But as soon as you shoot this stuff out over the open Internet it has to be re-resolved, it is a real pain.

Probably in most circumstance it will be really difficult to physically connect the datacenters. There are a couple of places, Texas I believe, where some of these people are actually in the same datacenter and there you could physically connect and literally run optical fiber between them all. But in most of these situations there are 50 miles away. But what you can do is you can optimize your routing and say: "I know that a lot of my customers want to talk to these other people. Let's start to create networks." And if I want to get to some Yahoo service all I need to do is get to the first Yahoo node and then let them use their optimized backhaul.

THORSTEN: Couldn't that discussion interfere with antitrust laws?

TOM: Not necessarily. It depends on what you are trying to achieve. It is not about excluding anybody. If you think about cloud as a service, if you think about Amazon and all the others: one of the problems that they have got now is that they are anti-competitive. It is really costly to switch and it is really difficult for these systems to interact with another. So you can't use BigTable with Amazon, it is completely and utterly infeasible. That is not necessarily a technical barrier.

THORSTEN: We have many clients where they start out with one business model and they are very happy with it, it is not that anything is broken. And it requires BigTable. But two years down the road, their business model changes. And suddenly they discover that something else might be actually better. But how do you get the data out, how do you change it, how do you make it backwards compatible? Are there API calls that you are required to do, is there the same communication paradigm or do you break the old API pattern? Very, very difficult.

TOM: The API issue is frustrating, but you can code your way out of it. You can't code your way out of getting your data out. That is a network problem. I actually think that cloud peering will create a market and reduce some of the risks for cloud clients. One of the things I would really like to see from the Yahoo side is to make it more possible for enterprises to work together, especially companies that want to use cloud to improve their core business but aren't interested in being the person that sells anything.

THORSTEN: Another problem my clients ask me a lot is about tiered service levels. You might only want to run the really important stuff yourself but everything else somewhere else. So you need to be able to peer to different kinds of clouds, different kinds of SLAs. That becomes difficult: How do I transport the data – maybe 2 terabytes – that I need to compute into this other facility?

TOM: Amazon lets you ship hard drives in and out. I don't know how they would feel like if you would ship in two terabytes with FedEx. Two terabytes would probably be ok, but more would probably be a problem. Yahoo's grids literally have petabytes of data on them. The load traffic of every Yahoo property gets dumped on to our grids and gets processed. From a regulatory point, particularly with 'Search', we have been really committed to how can we run our business in an efficient manner, while still being competitive on 'Search' per se and really pushing to help consumer's rights. Those two things seem at counterpoint. I want to reduce the amount of information that I am storing about my consumers for their privacy rights and for their benefit, but at the same time I want to run my business in a way where I can predict that I can help my advertisers and balance those needs.

What is really great is that, because we have got the grid, we dump all of that traffic on there, people work out the stuff they want and then you throw away the rest. You create a lot of reports that previously would have been impossible to do in any kind of real-time application. But instead of keeping the IP addresses indefinitely, as we would have done in the past, you can compute out the things you want – like location and all of these things – that are non-identifying and don't correlate the user to the data that you want. And you can compute all of those things out and do really complex stuff. But you can do it in real-time so that you are not actually accumulating even more data. It is feasible to do that and throw away the rest of it and say: 'We are not keeping the data, but we are keeping the report.'

THORSTEN: Which is an interesting problem and the reason why many telecoms don't throw away data. They don't know what is going to be interesting in one, two, or five years from now. So they feel they have to keep everything, beyond regulatory and legal compliance. Because if they do throw it away, who knows, maybe in five years this was the most valuable asset they ever had!

TOM: True, a difficult business problem. But technology will not help you there. It is an enabler. The fact that we keep lots of sensitive information in our grid – I am not going to get too specific, but I am sure you can guess – we can run reports that nobody would have dreamed of five years ago. It is all tradeoffs. As a search company, this was previously a huge issue. So we have to say 'It is the right decision for us to sacrifice some of the potential data mining we might want to do in the future'. That is the tradeoff, but it is a business tradeoff. Technology facilitates us being able to do those kinds of complex things.

THORSTEN: So where is that leading us in five years? What are we going to talk about in five to ten years?

TOM: One of our biggest customers is research. Yahoo Research loves all of the cloud services. The idea that you can have this available pool of resources beyond your normal day to day usage, I think that is going to become the pervasive trend in terms of development.

THORSTEN: In the future that is just how you use things.

TOM: it's just how you use things, right. We have seen that in our Continuous Integration[14]. It is really common now for teams to spin up servers, do a lot of heavy integration tests that, again, nobody would have done a few years ago. You have your Continuous Integrations servers and you have it like this. And now there are time schedules that say, let's buy a bunch of machines that require this much space on our internal cloud. This block of resources is going to be allocated against everybody who is doing the integration tests. And we are going to do some really heavy integration tests but we are going to do it with 500 machines. And suddenly people can do this type of testing they have always wanted to do and never could.

THORSTEN: It really accelerates your business and really also reduces risk factors.

TOM: This is the thing for us, this is what it's all about: Cost savings are nice and we can spread and manage the cost because we are doing across the biggest section of your business instead of buying hardware and piece ware. That is a benefit. But primarily what we are seeing is that this is really changing the way that developers can develop.

THORSTEN: So what do you think is going to come in 15 years? I know it is crystal-balling but…

---

[14] Continuous Integration (CI) is the strategy and practice of making sure that changes to a software project's code base are successfully built, tested, reported on, and rapidly made available to all parties after they are introduced.

TOM: It is crystal-balling… it's really tough. I would not be surprised if there is more consolidation. At some point somebody will solve the idea of renting the Cloud as a service. The problem is that there is a lot of regulatory issues there, a lot of compliance challenges, like PTI compliance[15]. At some point some of those problems are going to start to go away. And I think businesses will start to consolidate these things out to service providers. The technology is not there. Right now you see things like Opera mini, which is actually really fascinating because it says 'let's do stuff with a thin client.' And at some point even as we roll out technology, the cell phones themselves are hitting the same kind of barriers that laptops have hit or desktop computers have hit. You basically cannot pack anymore transistors into the device.

THORSTEN: And even if you do, it becomes too costly. The operational costs are too high. I am really interested in looking at Apple's iOS 4.0 and how multitasking might break applications, or even stall the whole phone, maybe.

TOM: And I think there are some interesting questions here. What I see more and more is: People will expect an experience, beyond what is possible with a single machine or device.

THORSTEN: Opera mini is a great example: virtualizing a browser can provide a very device-as well as network-customized experience.

TOM: I was with Jeff Bezos from Amazon and we were talking about velocity. He was freaking out about the fact that if you try to do something with Javascript and Opera Mini it sends it back to the server to do the Javascript and then sends you the result. And it is like 'Wow that is crazy'. But it makes sense, because the kind of devices where you really need Opera Mini – you don't really need Opera Mini on an iPhone, it is cute that they have gotten it on there – but the kind of devices where you really need Opera Mini are some Symbian 9300 or something…

THORSTEN: Right! And then suddenly, in general what you can achieve is also a load balancing and optimizing your battery performance. So maybe the lower your battery runs, the less communication you do and the more you do on-device. Of if that computation is taking too long or is too expensive maybe you do want to push the task out into the Cloud.

TOM: One of the constraints we have right now is network. Even if I have this beautiful phone with a big high resolution screen – like the iPhone 4's 960-by-640 pixel resolution at a mind boggling 326 pixel per inch, wow! – I can't send the graphical information to render that screen over current mobile networks. We will figure out network bandwidth, offloading to Wi-Fi, and some of those other things…

THORSTEN: … but screen size and capabilities will improve with it, too…

---

[15] PTI stands for Produce Traceability Initiative, the produce industry's focus to implement a common protocol for labeling all products at the case level for purposes of tracing back through the supply chain from retailer to farm. As a result, grower-packer-shippers of fresh produce are the ones tasked with the labeling step for each case of product they pack.

TOM: Exactly! But if both display and bandwidth improve, we might reach computational limits of a device of a certain size, the storage limit of the device of a certain size, etc....

THORSTEN: ... or simply network latency. There is a limit to the speed of light.

TOM: The expectations of the users, though, will be different. I personally expect most of this stuff to come with thin clients. If you have a device it will be: Here is a mostly thin client. It can probably do some old phone stuff if you are somewhere within the network. But in general it will be a way to access and render...

THORSTEN: ...you will try to avoid the on-device computation as much as possible...

TOM: ...right, because the on-device computation is then a much more degraded experience than it used to be, than it is right now. Even the desktop experience: You sit at a desk, but you don't want to have the power house... Maybe we will all have a home server in the basement, who knows. Maybe that power will be off in the Cloud or in a remote location, I don't know. I am assuming that at some point, in 15 years, when we can serve 100 gigabit connection to the home, you can just render everything in a server somewhere and just pump 100 frames a second to somebody's house.

THORSTEN: Tier 1 telecoms are usually involved in several of these projects. You can have an awesome interactive experience to a picture frame if it simply does MPEG 2 and has some feedback channel back into the telecom's head-end. OnLive is another great example, Cloud based gaming! Well, if you can do twitch gaming in the Cloud I'm sure you can do some word processing and other business applications in the Cloud. But the question is: what role does a telecom play in that game then? Will they be reduced to do IP transit? Is that actually a bad thing, to do IP transit really well? Is that what telecoms as a brand will stand for? What is Yahoo's or your expectation of telecoms, what service or product would you expect a telecom should offer?

TOM: This map here is actually really interesting (see map at http://www.cloud-peering.com) with regards to interconnection.

THORSTEN: Interconnection and actual light path routing and switching?

TOM: Yes. I was recently fixing a bug on something and the problem was that when a user connected to a different datacenter than the specific OAuth[16] application was connected to, the datacenter that the user was using did not know about the OAuth keys.

THORSTEN: Was that an issue with dynamic load balancing in the Cloud and IP or DNS address resolution?

---

[16] Open Authorization (OAuth) is an open standard that allows users to share their private resources (e.g. photos, videos, contact lists) stored on one site with another site without having to hand out their username and password.

Tom: It was a deployment defect, so we tried to pick some data centers that were really far away, we picked Dallas and Singapore. They are far away from another. But the database was actually routing so fast that we could not test the fix.

[Laughter]

The databases were tuned so well in the backhaul that we could not test our fix.

Thorsten: Interesting. So the only way to test it was to break something?

Tom: Yes, we actually had to slow it down. But the material point is this: The actual backhaul is like an incredibly important tool. Right now, we are coming up with all kinds of tricks and things that we have to do to improve the backhaul. That gets really costly and there are some really interesting problems there. And a backhaul is really expensive. It is crazy! Interconnectivity between continents is incredibly expensive. So for us trying to run a business, … maybe people are flying more, trying to create a social graph that is transcontinental is really difficult and really expensive. There is obviously cost involved in maintaining transatlantic lines but that is a really fascinating area where there are a lot of connectivity problems for us. One of the other things that is becoming increasingly important is geo-location. For example, if we have got a customer in Portland on the West Coast, we want to glue them to one of the datacenters either in California or in Seattle. What I have experienced is that telecoms are pretty bad at running things like DNS. Bad as it is, telecom's basic Internet services haven't been tip-top.

Thorsten: Which company or service is the benchmark there?

Tom: OpenDNS is great, the Google Public DNS service is great. But one of the problems is that if Google DNS has got a server on the East Coast, even though the user is currently staying at the West Coast we end up gluing them to a server on the East Coast by mistake, because when Google did the DNS look up we saw, 'Oh this is coming from an East Coast IP pool'. So there all of these strange problems where we, as a service provider, are trying to optimize our network topology to fit our end users and our customers. Nobody is really creating those relationships. So far nobody is stepping up and saying 'Look, we are going to try to be in charge, or we are going to try and lead in some way of good behavior about how you are going to describe this stuff.'

Thorsten: And you think that is a telecom play?

Tom: Well, the telecoms own the relationship with the end user.

Thorsten: Is that so? I am not sure about that, isn't it Yahoo who owns the relationship with the end user, as you aptly described in the beginning?

Tom: I think we own the end customer in the sense of people using Google DNS because telecoms' DNS sucks. Just think about this: When I was on a Web2.0 conference session and connected to the in-room Wi-Fi network the DNS server was geo-locating me in Sebastopol or something like that. So that is not useful to us, the granularity is not decisive. At some point the ISPs and the telecoms deliver the user the Internet and then after that the user is kind of alone.

THORSTEN: I see what you are saying. Telecoms often have large investments into vendors and networks. One could say that debt management is one of telecoms' core competencies. But how do you find digital products and services that fit into the up to now usual planning cycles of ten or more years – because you have those large infrastructure investments. Fiber-to-the-Home is a good example. Verizon's cost for every home passed was $1,800 in the beginning. Right now it is probably something around $800. But telecom networks are regulated and there are legal requirements in Germany, which also impacts telecoms' planning horizon.

TOM: There are ways that we want to connect with our customers. And our customers have to get to the Internet via the telecoms at the end of the day. And there is this big gap in the middle of mystery madness how the customer got from the ISP to Yahoo. If there is some way to short circuit that, there is potentially value there, value worth Yahoo spending money on. One of the things Google has been doing is, and this is pure speculation on my part but I think it is fairly good speculation, while doing Street View, they have been mapping wireless hotspots. So they are having this great geo-location database.

THORSTEN: This is just pure speculation on my part, but if just as Google could identifying hotspots, they could do pattern recognition of fiber and DSL street cabinets so they could understand planned or existing infrastructure roll outs.

TOM: That would be fascinating. A friend of mine is an engineer for Orange in the UK and we were talking about cell tower locations. Orange cell towers are running on software radio. And he was talking about whether they are going to draw a contingency plan, because there are a bunch of people mapping the location of cell phone radio towers in order to do triangulation. For example, the people who use Skycall, they have something like that. That would circumvent the value of Orange. The question is whether you prevent them for doing triangulation…

THORSTEN: … or providing it more cost effective.

TOM: And this is the point. In general, from our point of view, if you want to know where people are and they have a cell phone, AT&T or Verizon or whoever making their data available either directly themselves or through a 3rd party vendor, like Skyhook – Great! That means we can get reliable information. But that means that there needs to be a relationship between us and the telecoms.

THORSTEN: An interesting business offer for telecoms. I am pretty sure that you would spend money on that.

TOM: Well, I am not an executive, I can't say whether we would but…

THORSTEN: Is that the same for the DNS problem? Google is running its own DNS service because telecoms' DNS "sucks", as they say. If telecoms' DNS would be much better, and would be offered at less cost than it actually takes Google to reinvent the wheel and does really nice DNS, Google would probably be the first to say that: 'You know what: Now that telecoms DNS is fixed, I am going to use their service.'

TOM: I think there is definitely a bunch of services where there is a lot of potential. There is a lot of potential synergy and potential overlap of interests there. Because they are the same customers and it's kind of doing it in a way which is okay for the customer and doing it in a way which fits our current models. Just look at the usual configuration for Wi-Fi access points. So we have channels one through eleven on 802.11. Everybody uses one, six, and eleven – which is great, because these are the ones where there is no overlap in spectrum. If I am in San Francisco, I have massive overlap because I have so many of access points using the same channels one, six, and eleven. OfCom is the independent regulator and competition authority for the UK communications industries, and they have a great report on this. Bottom line: the saturation really adds up. And we can't fix that. We can't fix the users Wi-Fi.

THORSTEN: It is also not your core competency.

TOM: It is not our core competency. But we can look at that and we can make some assumptions. Telecoms sell 8Mbps broadband. But we can't make assumptions on that speed, because it turns out that the Wi-Fi routing gear that telecoms give for free to the customers are not smart enough.

THORSTEN: They are not managed.

TOM: At home I am actually using an Apple base station and in my neighborhood there are lots of people on channel one, lots of people on channel six, and a few people on channel eleven. The base station automatically selected channel 10 because it has the least interference.

THORSTEN: This is the same thing that Aruba Networks is really good at – large complex roll outs. It is amazing what they do. They manage like sixteen different vendors better than the vendors themselves.

TOM: We have Aruba in our campus and it does crazy stuff and suppresses rogue wireless networks. If you would create a wireless network it would flood it with packets and shut it down. Coming back to telecoms: we don't control the end to end quality. Telecoms have an interest to ensure a good wireless experience. We as ISPs have also a vested interest in seeing that the user is getting the best of it at the home. And that's why it's interesting of Google to dip their toes into selling broadband, but I have never actually seen them address this specific issue.

THORSTEN: What would be interesting to see is whether Yahoo's brand is harmed if the wireless access point is wrongly configured. Would people say that Netflix's video on demand service sucks if they know that they have an 8Mbps Internet connection but the videos are blocky and grainy? How do I as a consumer know that this is a faulty network access configuration and not a faulty Internet-based service?

TOM: That is exactly the reason why I am arguing for a partnership with telecoms. In the future we will share many services, many networks, and many interests. Just as we share the consumers' mind and wallet. Services are going to be produced in the Cloud by many parties, and also transferred between partners, often without the knowledge of the end user. They will not care as long as their personal, financial, and legal boundaries are not violated…

THORSTEN: … or if they see an inherent benefit to sharing all sorts of information. Like Facebook.

TOM: Correct. But we need to get a better understanding and cooperation on many levels of the service production, end-to-end, through the Cloud. So that neither your, nor my brand is harmed, neither telecoms' nor Yahoo's service and revenues get impaired.

THORSTEN: Thank you so much for your time and insights.

TOM: You bet!

Appendix

Cited Work

Coleman, P., & Papp, R. (2006). Strategic alignment: analysis of perspectives. *2006 Southern Association for Information Systems Conference.*

Enhanced Telecom Operations Map. (2010, January 27). Retrieved March 24, 2010, from Wikipedia: http://en.wikipedia.org/wiki/Enhanced_Telecom_Operations_Map

Hagel, J., & Singer, M. (1999, March 1). Unbundling the Corporation. *Harvard Business Review* .

Henderson, J. C., & Venkatraman, N. (1993). Strategic alignment: leveraging information technology for transforming organizations. *IBM Systems Journal , 32* (1).

Nafus, D. (2010). An Anthropologist's Eye for the Tech Guy: Emerging Market Opportunities in a Post- BRIC World. *eComm 2010.* San Francisco.

Nafus, D., Howard, P., & Anderson, K. (2009). *Getting Beyond the Usual Suspects: Policy and Culture in ICT Adoption.* Intel.

O'Reilly, T. (2005, September 9). *What is Web 2.0.* Retrieved March 24, 2010, from O'Reilly: http://oreilly.com/web2/archive/what-is-web-20.html

About The Authors

THORSTEN CLAUS is a telco2.0 enthusiast, angel investor and adviser, creating startups for cloud services and telecom enablers, and supporting community and public healthcare projects. As a Managing Consultant at Deutsche Telekom Group's technology and management consultancy Thorsten is helping his international telecom and Internet clients to engineer product and service innovations and to create and execute strategies for a flat and free world. He has over fifteen years of telecom and information technology industry experience and was previously a senior network architect for T-Systems North America, responsible for creating blueprints and evaluating emerging technologies and architectures for telecom-grade billing systems, next generation operation support systems, fixed-mobile converged carrier infrastructures, and IPTV networks. He has a Master of Science in Information Technology and a Bachelor of Science in Business and Accounting from TU Darmstadt, and lives with his wife and two kids in Berkeley, California.

DANIEL KELLMEREIT has spent more than a decade helping large global companies to develop business strategies, innovate and shape markets, adopt emerging technologies, design strategic alliances and launch new products and services. He also supports early-stage companies to build compelling products, grow their business, and raise capital.

In his current position as CEO of Detecon Inc., a leading global management and technology consultancy, he is responsible for the North and South America region. Prior to this role, Daniel led the San Francisco office of Detecon Inc. and the Strategy & Innovation practice in the US. He works with clients in the telecom, internet, hardware, software and services industries. Prior to joining Detecon, Daniel was co-founder and CTO of a "software as a service" company, building products for the supply chain industry. He holds an MBA from Northwestern University's Kellogg School of Management.

YASMIN NARIELVALA is a Managing Consultant with Detecon Inc. She is responsible for leading strategic technology assessments for telecoms industry clients, scouting and assessing emerging technologies, services, and products. She is assisting in development of technology, business, and product strategies, and with more than 12 years of telecoms experience working for a diverse range of companies from Telstra in Australia, to Siemens in England, to a VoIP start-up in the early days (unfortunately not Skype), Yasmin has a passion for the industry and its wealth of opportunities to improve the future for all of us. She has a Masters of Science in Commerce and a Bachelor of Science in Engineering, and is based in San Francisco, California.

Contact Us – We Need Your Help!

We would like to hear from you! The interviews included in this book took place over the course of nine months. Many things happened since we began. People changed jobs and positions, companies were acquired, others failed and went out of business.

We set out to gain insights and intelligence on a long-term roadmap for telecom technology, product, and service innovations. As always with long-term outlooks, we probably got a few things wrong (hopefully only a few). Please tell us!

We want to add another chapter to this book where you – the reader – get your voice. To tell us where we were right; where things are more complex and layered than we thought; and where we were plain-out wrong.

But we need your help! Call us! Email us! Leave a comment on our website! Contact us on LinkedIn, Xing, Twitter, Facebook, etc.! We are looking forward to hearing from you.

WEBSITE:	http://www.thefutureofcloud.com
MAILING ADDRESS:	Detecon, Inc. 128 Spear Street, 4th Floor San Francisco, CA 94105 USA
THORSTEN:	thorsten.claus@detecon.com +1-415-830-4161
DANIEL:	daniel.kellmereit@detecon.com +1-415-904-7979
YASMIN:	yasmin.narielvala@detecon.com +1-415-904-7907

Made in the USA
Lexington, KY
14 October 2010